Monographs in Electrical and Electronic Engineering

Editors: P. HAMMOND, D. WALSH

Monographs in Electrical and Electronic Engineering

R. L. Bell: *Negative electron affinity devices* (1973)
H. C. Wright: *Infrared techniques* (1973)

Avalanche-diode microwave oscillators

G. GIBBONS

CLARENDON PRESS · OXFORD
1973

Oxford University Press, Ely House, London W. 1

GLASGOW NEW YORK TORONTO MELBOURNE WELLINGTON
CAPE TOWN IBADAN NAIROBI DAR ES SALAAM LUSAKA ADDIS ABABA
DELHI BOMBAY CALCUTTA MADRAS KARACHI LAHORE DACCA
KUALA LUMPUR SINGAPORE HONG KONG TOKYO

Printed in Great Britain
at the Pitman Press, Bath

Preface

This book has been written primarily for undergraduates specializing to some extent in their basic course or taking optional topics in modern electronic engineering. It is intended also to be of value to graduate students or graduates in industry who are working in related fields and require an introductory treatment of the theory and performance capability of the avalanche-diode oscillator.

Wherever possible simple physical models have been used to elucidate the device physics and detailed mathematical analysis has been kept to a minimum. The emphasis has been placed on the basic principles of operation of the oscillator and on the mechanisms which set limitations on its microwave performance. Practical aspects of the device technology are discussed, including fabrication techniques and the relative merits of different semiconducting materials.

During the past eight years I have benefited greatly from discussions with many former colleagues at Bell Telephone Laboratories and at the Sperry Rand Research Center as well as present associates at the Allen Clark Research Centre. I am particularly indebted to S. R. Geraghty of University College London for permission to use some of his unpublished work and for many valuable discussions. I am indebted also to Dr. D. Walsh of Oxford University for his continued encouragement and for suggesting several improvements in the manuscript. I would like to thank Dr. J. J. Purcell for undertaking the rather laborious task of proof reading. Finally it goes without saying that any book describing the results of recent research and development depends for its existence on the work of many researchers in the field.

G. Gibbons

Allen Clark Research Centre,
The Plessey Company Limited,
Caswell, Towcester, Northants.
February, 1973

Contents

1. INTRODUCTION 1
References 3

2. HIGH-FIELD EFFECTS IN SEMICONDUCTORS 4
2.1 Introduction 4
2.2 Drift velocity and mobility 5
2.3 Impact ionization and avalanche multiplication
– theory of impact ionization, avalanche multiplication 12
References 21

3. REVERSE-BIASED p–n JUNCTION 23
3.1 Voltage-current characteristic 23
3.2 The depletion-layer approximation 26
3.3 The avalanche breakdown condition 28
3.4 Avalanche breakdown of IMPATT and TRAPATT diodes
– one-sided abrupt junctions, symmetrical-step junctions, the Read diode, TRAPATT diodes 30
3.5 Space-charge resistance and temperature effects 40
References 43

4. IMPATT MODE – PRINCIPLES OF OPERATION 44
4.1 Introduction 44
4.2 Transit-time delay 44
4.3 Avalanche delay 46
4.4 Oscillator efficiency 48
4.5 Small-signal impedance
– avalanche-zone impedance, the drift-zone impedance 49
4.6 Space-charge effects 53
4.7 Generalized small-signal analysis
– uniformly avalanching region, multiple uniform-layer approximation 56
4.8 Large-signal analysis 62
References 64

5. PERFORMANCE LIMITATIONS OF IMPATT OSCILLATORS 66
5.1 Inherent material factors affecting efficiency
– effect of ionization rates, efficiency fall-off at high frequencies, minority-carrier storage, parasitic resistance 66
5.2 Transit-time limits 76
5.3 Thermal limitations 78
5.4 Power–frequency limits for abrupt-junction IMPATTs 81
References 86

6. PRACTICAL IMPATT OSCILLATORS 88
6.1 Device fabrication
– diode construction techniques, thermal impedance, contacts, special fabrication techniques for high-frequency diodes 88
6.2 IMPATT oscillator circuits
– practical circuits and stability, circuits for millimetre-wave IMPATTs 97
6.3 Performance of IMPATT oscillators
– power, efficiency, and frequency 103
References 106

7. TRAPATT OSCILLATORS 108
7.1 Mode of operation 108
7.2 The avalanche shock front and trapped-plasma formation 109
7.3 Plasma extraction 111
7.4 Triggering mechanisms 114
7.5 TRAPATT device theory
– plasma region, recovery region, frequency of operation, power–frequency relation 115
7.6 Practical device design 121
7.7 High-frequency TRAPATTs 125
7.8 TRAPATT circuits 127
7.9 Harmonic extraction 129
7.10 Practical results 131
References 133

INDEX 135

List of principal symbols

C = junction capacitance
D_p = hole diffusion coefficient
E = electric field
E_B = breakdown field
E_D = electric field in drift space
E_g = forbidden energy gap
$\mathscr{E}_a$ = a.c. field amplitude
F = punch through factor
f_D = drift frequency
g = oscillation growth rate
i_p = a.c. particle current
i_d = displacement current
J = current density
J_s = saturation current density
j_a = a.c. conduction current density
j_d = displacement current density
j_n = electron current density
j_p = hole current density
K = thermal conductivity
k = Boltzmann's constant
l = mean free path for acoustic-phonon scattering
l_i = mean free path for ionization
l_{op} = mean free path for optical-phonon scattering
M_n = electron multiplication factor
M_p = hole multiplication factor
m^* = effective mass
N = impurity density
N_A = acceptor density
N_D = donor density
n = electron density
n_i = intrinsic carrier density
n_p = electron density in trapped plasma

n_R = electron density in TRAPATT recovery region
P_o = output power
p = hole density
p_P = hole density in trapped plasma
p_R = hole density in TRAPATT recovery region
Q_d = diode Q factor
q = electronic charge
R_p = parasitic resistance
R_{sc} = space-charge resistance
r = diode radius
T = absolute temperature
T_e = electron temperature
T_B = burn out temperature
U = shock front velocity
u = velocity of sound
V = voltage
V_A = voltage drop in avalanche zone
V_B = breakdown voltage
V_{BO} = room temperature breakdown voltage
V_b = built-in voltage
V_D = voltage drop in drift zone
V_p = punch through voltage
v = carrier drift velocity
v_p = carrier velocity in trapped plasma
v_s = saturated drift velocity
w = depletion-layer width
w_B = depletion-layer width at breakdown
X_c = circuit impedance
x_A = avalanche-zone width
x_D = drift-zone width
Z = total diode impedance
Z_A = avalanche-zone impedance
Z_D = drift-zone impedance

α = electron ionization rate
β = hole ionization rate
β_0 = temperature coefficient of breakdown voltage
ϵ = permittivity
ϵ_i = ionization energy
ϵ_{op} = optical phonon energy

ϵ_T = electron energy
η = conversion efficiency
θ = thermal impedance
θ_0 = transit angle
μ = mobility
μ_0 = low-field mobility
μ_I = mobility due to ionized impurity scattering
μ_n = electron mobility
μ_p = mobility due to phonon scattering
μ_p = hole mobility
ρ = resistivity
τ = transit time
τ_0 = relaxation time
τ_A = avalanche shock front transit time
τ_E = plasma extraction time
τ_{op} = relaxation time for optical-phonon scattering
τ_p = lifetime for hole recombination
ϕ_{ms} = metal-semiconductor work function
ω = angular frequency
ω_a = avalanche resonance frequency

1 Introduction

The development of solid-state microwave generators has been one of the most rapidly expanding areas of semiconductor device research over the past ten years. The motivation has been the significant cost savings in microwave systems and the consequent extensions in the applications of microwaves that would result if a simple, cheap, and reliable source were available. Several devices have now been developed that are capable of generating microwave power at levels adequate for many applications in microwave systems and, of these, the avalanche diode and Gunn oscillators are perhaps the most significant. Both are two-terminal negative-resistance devices that are simple to make and are quite compatible with microwave integrated-circuit technology. The Gunn oscillator is a bulk-effect device in which the mechanism giving rise to the negative resistance is a basic property of the semiconducting material used in its construction. Not all semiconductors possess the required property: n-type GaAs has been used almost exclusively for this type of oscillator although other III-V compounds also exhibit the bulk negative-resistance effect.

The avalanche diode differs from the Gunn device in that the mechanism giving rise to the negative resistance is a property of the p–n junction. Thus any semiconductor in which it is possible to fabricate a junction can be used, and these oscillators have been constructed in Si, Ge, and GaAs. The working principles of the device were first described by Read [1] in 1958. However, the idea of obtaining a negative resistance from a reverse-biased p–n junction dates back to an earlier paper (1954) by Shockley [2], in which he showed that by injecting an electron bunch into the depletion layer of a reverse-biased p–n junction a 'transit-time negative resistance' is produced as the electrons drift across the high-field region. Shockley suggested a forward-biased junction as the injecting cathode. Unfortunately the negative resistance of this device is small and the microwave power output low. Read showed that an improved negative resistance is obtained when impact ionization is used to inject the electrons. By exploiting the time delay required to build up an avalanche discharge by impact ionization coupled with Shockley's transit-time delay, he showed that an efficient microwave oscillator could be constructed. Read proposed a p^+nin^+ diode specifically designed to localize the injecting avalanche region at the p^+–n interface. This structure proved difficult to make; it was not until 1965 [3] that the first experimental Read diodes were fabricated, and these were low-frequency and low-power devices. At about the same time, Johnson, De Loach, and Cohen [4] at

Bell Telephone Laboratories demonstrated that significant amounts of microwave power could be obtained using simple computer diodes. This was a significant result, showing that the complicated Read structure was not required. Since then, significant advances have been made both in the device technology and in the understanding of the oscillation mechanism. The mode of oscillation conceived by Read has been called the IMPATT mode (for IMPact-Avalanche Transit Time). At the present time (1972) experimental devices have produced powers of five to ten watts in continuous operation and frequencies over 100 GHz have been obtained. At X band, devices producing one watt are now commercially available from several manufacturers. Several years after the first IMPATT oscillators were constructed a new mode of oscillation in the avalanche diode was discovered by Prager, Chang, and Weisbrod [5] at R.C.A. This mode differed distinctly from the operation described by Read, in that the oscillation frequency was well below the transit-time frequency and the conversion efficiency was about twice the maximum value calculated by Read. It was shown some time later that the mechanism producing the high-efficiency oscillation involved a new transient breakdown effect in a p–n junction. This mode has been called the TRAPATT mode (for TRApped–Plasma Avalanche-Triggered Transit).

The IMPATT and TRAPATT modes can be broadly distinguished into two classes of oscillator. The IMPATT oscillator is best suited for low-power (one to ten watts) high-frequency (up to over 100 GHz) continuous working applications. The TRAPATT oscillator is best suited for high-power, high-efficiency pulsed applications at low frequencies (up to about ten GHz). The power levels presently available in the IMPATT mode are adequate for the primary sources in microwave relay and trunk waveguide communication systems. At millimetre-wave frequencies ($>$ 30 GHz) IMPATT oscillators are being used as pump sources in parametric amplifiers. Many systems require low-noise receivers incorporating a parametric amplifier, and a large portion of the cost of such an amplifier is in the high-frequency pump source. It has been estimated that significant cost savings can be obtained if an IMPATT oscillator is used as the pump in place of the conventional klystron or solid-state multiplier chain. Pulsed powers at the kilowatt level have been produced by TRAPATT oscillators at frequencies in the range 1–2 GHz. These devices do not compete with IMPATTs for the same applications, but are likely to find widespread use where high efficiency and high power are required but continuous operation is not essential. One such application, in which high-power solid-state oscillators appear particularly attractive, is the phased-array radar system composed of an active array of solid-state sources. In this system the microwave beam is electronically scanned by externally controlling the phase of each element in the array.

The avalanche-diode oscillator is a 'hot-electron' device and its properties are determined by the non-ohmic behaviour of electrons and holes at high values of electric field. In Chapters 2 and 3 these high-field effects are reviewed and the concepts of mobility, saturated drift velocity, impact ionization, charge multiplication, and avalanche breakdown are introduced. The treatment here does not include a detailed discussion of the electron-scattering processes, but is intended only to introduce the basic ideas which are essential to an understanding of the mechanisms producing the negative resistance in the avalanche diode. In fact, a full discussion of electron scattering at high fields is, in itself, a topic worthy of a book, and the reader who wishes a more profound analysis of these effects is referred to the works of Moll [6] and Conwell [7]. In the last four chapters, which constitute the main part of the book, the theoretical and practical aspects of IMPATT and TRAPATT oscillators are presented. In analysing the devices we have relied as much as possible on analytical methods, although simulation and numerical techniques have been used quite extensively in the literature. This approach has been chosen deliberately because it emphasizes the relevant physical principles involved though, in some cases, at the expense of treating simplified device models. This is seldom a serious limitation, but where it is, references have been given to more exact analyses of more realistic models.

References

1. Read, W. T. A proposed high-frequency negative-resistance diode. *Bell Syst. Tec. J.* **37**, 401 (1958).
2. Shockley, W. Negative resistance arising from transit time in semiconductor diodes. *Bell Syst. tech. J.* **33**, 799 (1954).
3. Lee, C. A., Batdorf, R. L., Wiegmann, W., and Kaminsky, G., The Read diode – an avalanching, transit-time negative resistance oscillator. *Appl. Phys. Lett.* **6**, 89 (1965).
4. Johnson, R. L., DeLoach, B. C., and Cohen, B. G. A silicon-diode microwave oscillator. *Bell Syst. tech. J.* **44**, 369 (1965).
5. Prager, H. J., Chang, K. K. N., and Weisbrod, J. High-power, high-efficiency silicon avalanche diodes at ultra-high frequencies. *Proc. IEEE.* **55**, 586 (1968).
6. Moll, J. L. *Physics of semiconductors* McGraw-Hill, New York (1964).
7. Conwell, E. M. *High-field transport in semiconductors.* Academic Press, New York (1967).

2 High-field effects in semiconductors

2.1 Introduction

In semiconducting materials departures from Ohm's law are frequently encountered. This is in contrast with metals, where, in general, the linear relationship between current and voltage is obeyed. This fact was first pointed out by Shockley [1] on the basis that the average electron energy in a semiconductor is quite small ($\frac{3}{2}\ kT$ compared with several eV in a metal). As a result, small increases in energy represent large deviations from thermal equilibrium for electrons in semiconductors. In many cases deviations from Ohm's law result from changes in carrier density produced by thermal effects, surface effects, and space-charge effects. Here, however, we are concerned with a bulk phenomenon that produces a deviation from the linear relationship between current and voltage. This occurs when the semiconductor is subjected to a large electric field which accelerates the electrons to velocities greater than the velocity of sound in the crystal. The low-energy acoustic-phonon scattering by which the electrons lose their energy then becomes an inefficient process and the electrons gain energy from the field faster than they can lose it to the lattice. The effective temperature characterizing the electron velocity distribution rises above the lattice temperature and the carriers are called 'hot electrons'. When the electrons are no longer in thermal equilibrium with the lattice, additional mechanisms for energy loss become important. In particular the electrons may lose energy by the excitation of optical phonons. Optical-phonon emission is a very efficient means of energy transfer to the lattice and causes the average drift velocity of the electron to reach a limiting value which is independent of field. When this saturated drift velocity is reached, the current through the sample will not respond to changes in voltage across it, provided the electric field within the semiconductor bulk does not drop below the value required to maintain the constant velocity. We shall show, in Chapter 4, that this effect can be used to develop a phase shift between current and voltage when electrons are in transit through the depletion region of a p–n junction. This is the first of two basic phenomena responsible for the negative resistance of the avalanche diode.

At even higher fields another mechanism for energy loss becomes important. This is the process of impact ionization, whereby an energetic electron suffers a collision with an atom of the crystal and imparts sufficient energy to a valence electron so that it becomes free. That is to say, an electron is raised from the valence band into the conduction band, leaving behind a hole in the valence band. Thus, as a result of one such collision, a hole–electron pair is created. The energy required by an electron to produce an ionizing collision, the ionization energy ϵ_i, must clearly be greater than the band-gap energy E_g of the semiconductor. In fact, since momentum must be conserved in the collision, and if we assume that the incident electron and the created hole and electron all have the same effective mass, then the ionization energy is $\geqslant \frac{3}{2} E_g$. This can be shown by applying the laws of conservation of energy and momentum to the minimum-energy case where all three particles move away with equal velocity after the collision. For an electron to acquire this energy it must travel a large distance in the lattice without suffering any optical-phonon collisions, and although the probability of this happening is generally quite small, it becomes appreciable at large values of applied field. The hole and the electron created by the collision are accelerated in opposite directions under the influence of the electric field and they can in turn suffer ionizing collisions, each producing one new hole–electron pair. In this way the free charge, and therefore the particle current, will build up by the process of impact-avalanche multiplication. The current produced by the charge multiplication requires a finite time to develop during the avalanche build-up, and it is the time delay inherent in this process which contributes the second basic phenomenon necessary to an understanding of the origin of the negative resistance in the avalanche diode.

The process of avalanche multiplication will be described in more detail in §2.3, but first the relationship between drift velocity and electric field for electrons in semiconductors will be discussed very briefly. Although electrons will be referred to specifically, it should be understood that holes behave in a similar manner.

2.2 Drift velocity and mobility

When an electric field E is applied to an n-type semiconductor the electrons are accelerated in the direction of the field and an electric current will build up. However, the increased electron velocity will cause the electrons to collide more frequently with imperfections in the crystal. These collisions cause the current to decay, and it is convenient to define a time constant τ_0, proportional to the mean free path and inversely proportional to the random electron velocity, which describes the rate of relaxation of the disturbance. This time constant depends, in general, on the kinetic energy of the electron, but let us consider the simple case where τ_0 is the same for all electrons. Then the decay of current is given by

$$\frac{\mathrm{d}I}{\mathrm{d}t} = -\frac{I}{\tau_0}. \qquad (2.1)$$

A steady state is obtained when the energy gained from the field is exactly balanced by the energy lost due to collisions. The rate of increase of the current is given by

$$\frac{\mathrm{d}I}{\mathrm{d}t} = qn\frac{\mathrm{d}v}{\mathrm{d}t} = \frac{q^2nE}{m^*}, \tag{2.2}$$

where m^* is the effective mass of the electrons. From eqns (2.1) and (2.2) the steady-state current is

$$I = \frac{q^2n\tau_0E}{m^*}, \tag{2.3}$$

and using $I = qn_v$, where v is the average drift velocity imparted to the electrons, we obtain a relationship between the velocity and the field

$$v = \frac{q\tau_0E}{m^*} = \mu E, \tag{2.4}$$

which defines the electron mobility μ in terms of the charge, effective mass, and relaxation time. The physical details of the scattering processes are effectively described by the relaxation time, which, as we have already noted, depends on the electron velocity. When the velocity dependence of τ_0 is taken into account the mobility is given [2] by

$$\mu = \frac{q\langle v^2\tau_0\rangle}{m^*\langle v^2\rangle}, \tag{2.5}$$

where the averages are taken over the velocity distribution.

Different types of crystal imperfections play a role in the scattering processes depending on the kinetic energy of the electrons. It is important to realize that we need not consider collisions between electrons and the crystal atoms if these are located in their equilibrium positions. The effect of the periodic lattice potential is contained in the electron effective mass, and only departures from periodicity such as thermal vibrations of the atoms or ionized impurities act as scattering centres.

For a more detailed discussion of the scattering processes, which is beyond the scope of this book, the reader is referred to the work of Moll [2]. We will, however, present expressions for the mobility at low, intermediate, and high values of electric field and show that the drift velocity reaches a limiting value at high fields.

Low fields

At low fields the principal scattering interactions are with acoustic phonons and with ionized impurities. In the former case, assuming a constant mean free path and isotropic scattering, the low-field mobility from eqn. (2.5) becomes [2]

$$\mu_{\mathrm{p}} = \frac{4ql}{3(2\pi m^*kT)^{\frac{1}{2}}}, \tag{2.6}$$

where T is the lattice temperature and l is the mean free path for acoustic-phonon scattering. The thermal vibrations of the crystal are more intense at high temperatures and so, for this type of scattering, the mobility decreases as $1/T$. Scattering by ionized impurities is coulomb scattering and is most effective on low-velocity electrons. Impurity scattering is not isotropic and the dependence of τ_0 on v is quite complex. Consequently the average quantities in (2.5) are very difficult to evaluate and approximations have to be made. The mobility has been calculated by Conwell and Weisskopf [3] from eqn (2.5), obtaining a result of the form

$$\mu_{\mathrm{I}} = \frac{A\epsilon^2 T^{\frac{3}{2}}}{Nm^{*\frac{1}{2}}}\left[\ln\left(1+\frac{BT}{N^{\frac{1}{3}}}\right)^2\right]^{-1}. \tag{2.7}$$

The agreement between experimental results and the Conwell–Weisskopf formula is reasonably good, but the theory somewhat overestimates the mobility. Since ionized-impurity scattering is most effective for electrons with low velocity, this mobility increases as the temperature is increased. As a result the net electron mobility is limited by ionized-impurity scattering when the temperature is low or when the impurity density is high, and by acoustic-phonon scattering when the temperature is high or the impurity density low. Over a wide range of temperatures and impurity densities, both scattering mechanisms contribute to the electron mobility. In such cases the evaluation of the $\langle v^2\tau_0\rangle$ term in eqn (2.5) is difficult, owing to the complexity of the relationship between the decay time and the velocity. A simple approximation is obtained by adding the reciprocal of the two mobilities,

$$\mu_0 \simeq \frac{\mu_{\mathrm{p}}\mu_{\mathrm{I}}}{\mu_{\mathrm{p}}+\mu_{\mathrm{I}}}, \tag{2.8}$$

but this is accurate only to within about 30 per cent.

The measured values of the low-field mobility for holes and electrons in Ge, Si, and GaAs at room temperature are shown as a function of the ionized impurity concentration in Fig. 2.1 [4] In all three cases the electron mobility is higher than the hole mobility corresponding to the lower effective mass of the electron in these semiconductors. At low impurity concentrations the mobility is constant, as expected from eqn (2.6). However, at high concentrations the mobility approaches the impurity-scattering limit, which decreases roughly proportional to N^{-1}. The resistivity ρ of the sample can be calculated from

$$\rho = q\mu_{\mathrm{n}}n + q\mu_{\mathrm{p}}p, \tag{2.9}$$

where μ_{n} and μ_{p} are the low-field mobilities of electrons and holes respectively. The resistivities at room temperature for n-and p-type Ge, Si, and GaAs are shown in Fig. 2.2 [4]

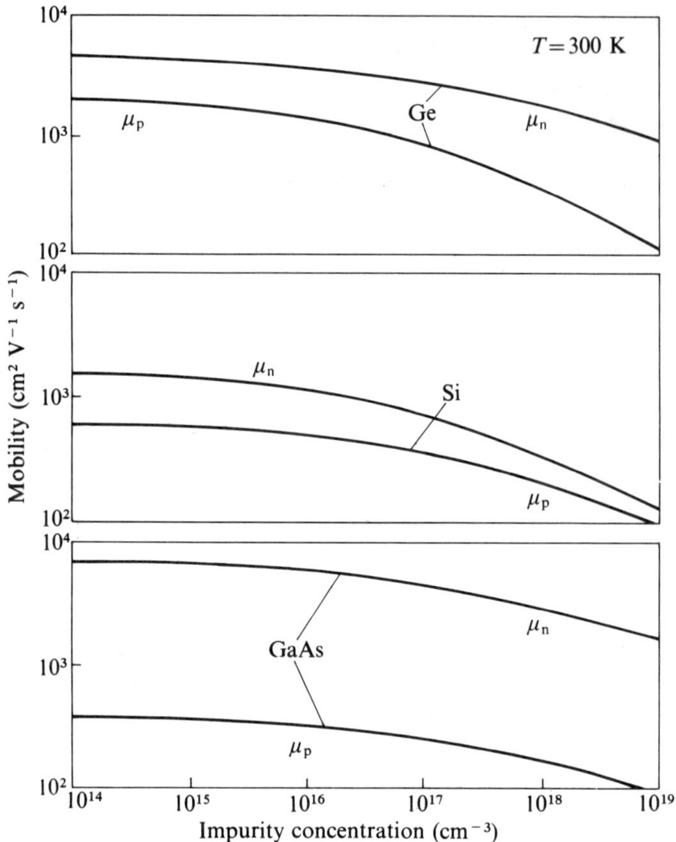

Fig. 2.1. Low-field mobilities of n- and p-type Ge, Si, and GaAs at 300 K versus impurity concentration (after Sze and Irvin [4]).

Intermediate Fields

The acoustic-phonon scattering is not an efficient means for the electrons to lose their energy, and as the field is increased the effective temperature T_e characterizing the electron velocity distribution increases above the lattice temperature T. In the intermediate-field range the electron mobility is given by an expression similar to (2.6) with the temperature T replaced by the hot-electron temperature T_e

$$\mu = \frac{4ql}{3(2\pi m^* kT_e)^{\frac{1}{2}}}. \tag{2.10}$$

By equating the rate of energy loss of the hot electrons to the rate of energy

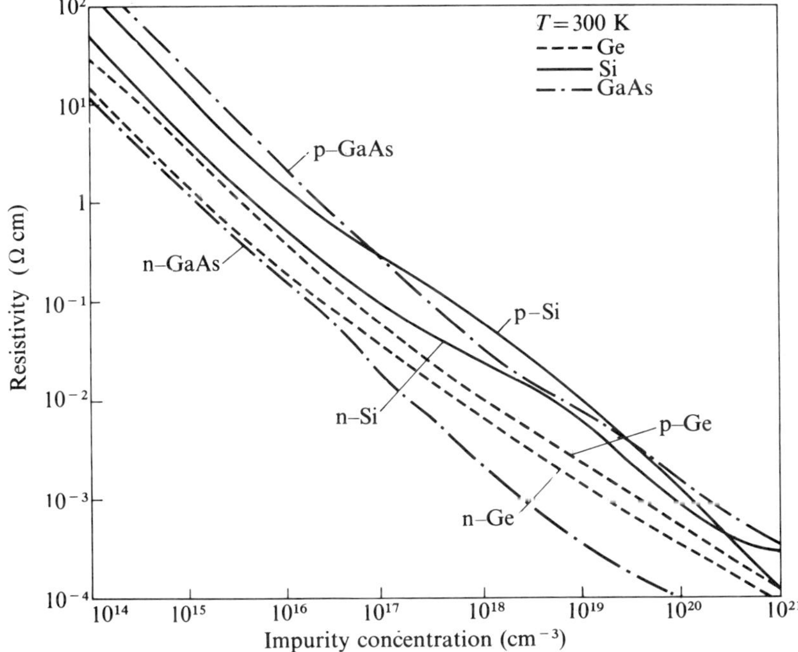

Fig. 2.2 Resistivity versus impurity concentration for Ge, Si, and GaAs at 300 K (after Sze and Irvin [4]).

gain from the field (qEv), Shockley [1] has derived the following relationship between T_e and E:

$$\left(\frac{T_e}{T}\right)^2 - \left(\frac{T_e}{T}\right) = \frac{3\pi}{32}\left(\frac{\mu_0 E}{u}\right)^2, \tag{2.11}$$

where u is the velocity of sound in the crystal. When the electron velocity ($\mu_0 E$) is small compared with u the electrons are in thermal equilibrium with the lattice. This defines the maximum value of field at which the low-field approximation and Ohm's law apply. In the case of Ge the velocity of sound is 5×10^5 cm s^{-1}, and for fields less than about 10^4 V cm^{-1} the low-field mobility is obtained. For fields above the low-field limiting value and for small departures of the electron temperature from the lattice temperature, (T_e/T) is given by

$$\left(\frac{T_e}{T}\right) = \left(\frac{3\pi}{32}\right)^{\frac{1}{2}} \frac{\mu_0 E}{u}. \tag{2.12}$$

Thus from eqns (2.10) and (2.12) the mobility in the intermediate field range

is proportional to $1/E^{\frac{1}{2}}$. Therefore the drift velocity and the current are proportional to $E^{\frac{1}{2}}$ and not E. This departure from linearity marks the beginning of saturation in the velocity–field characteristic and we shall show that at even higher fields the velocity tends to a limiting value.

High Fields

The measured drift-velocity versus field curves for both holes and electrons in Ge, Si, and GaAs at room temperature are shown in Fig. 2.3. [5–11] In all cases the linear relation between drift velocity and electric field is valid up to fields in the range 10^3 V cm^{-1} to 10^4 V cm^{-1}. As the field is increased, significant departures from the linear relation are apparent, as expected, and at high fields the velocity tends to saturate in all cases. To explain the saturation effect, scattering by optical phonons must be included. Optical phonons are high-frequency thermal vibrations of the lattice in which the two face-centred cubic sublattices of the crystal are vibrating in opposite directions. These modes are not excited at room temperature if $kT \ll \epsilon_{op}$, the characteristic optical-phonon

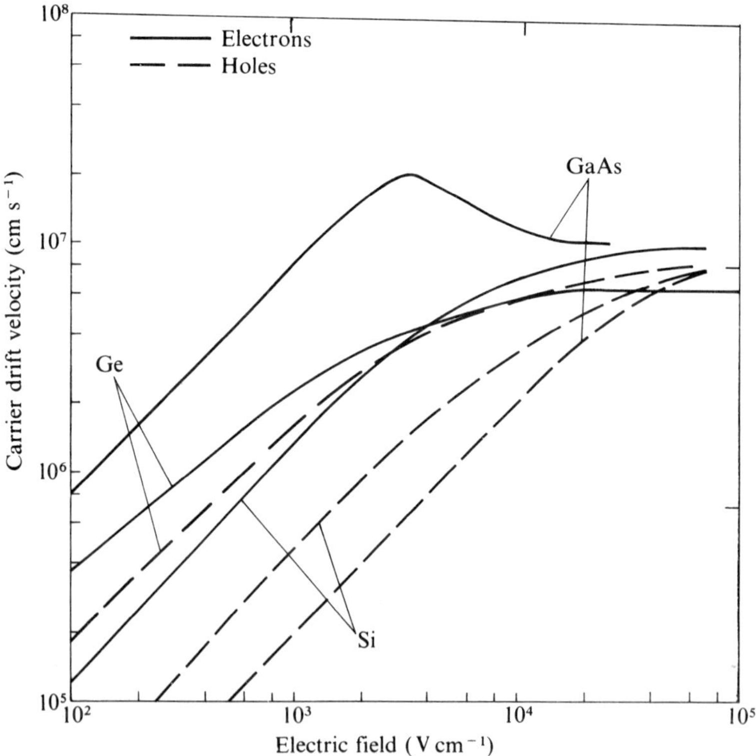

Fig. 2.3 Drift velocities of electrons and holes at 300 K in Ge, Si, and GaAs as a function of electric field [5–11].

energy. However, when an electron acquires an energy greater than ϵ_{op} the probability of this electron losing a large fraction of its energy by the emission of an optical phonon becomes appreciable. The optical-phonon energies for Ge, Si, and GaAs are given in Table 2.1. [12]

We can define a relaxation time τ_{op} for optical-phonon scattering, similar to the acoustic-phonon case. The mobility in the high-field range is then given by

$$\mu = \frac{q\tau_{op}}{m^*}. \tag{2.13}$$

In each collision the electron loses an energy ϵ_{op}, and so the rate of energy loss by optical-phonon emission is

$$\frac{d\epsilon}{dt} = -\frac{\epsilon_{op}}{\tau_{op}}. \tag{2.14}$$

The electron gains energy from the field at a rate

$$\frac{d\epsilon}{dt} = (q\mu E)E, \tag{2.15}$$

which on average must balance the energy lost in optical-phonon emission. Hence from eqns (2.14) and (2.15) we get

$$\tau_{op} = \frac{(\epsilon_{op} m^*)^{\frac{1}{2}}}{qE}, \tag{2.16}$$

and the high-field mobility is

$$\mu = \left(\frac{\epsilon_{op}}{m^*}\right)^{\frac{1}{2}} \frac{1}{E}. \tag{2.17}$$

That is, the mobility is proportional to $1/E$ and the drift velocity is independent of field and given by

$$v_s = \left(\frac{\epsilon_{op}}{m^*}\right)^{\frac{1}{2}}. \tag{2.18}$$

We shall see in Chapter 5 that a large value of saturated drift velocity is a desirable material's property for avalance-diode oscillators. This is obtained when the optical-phonon energy is large and the carrier effective mass is small.

Before leaving this discussion of carrier drift velocity and mobility, some mention must be made of the electron velocity–field characteristic in GaAs which, as shown in Fig. 2.3, possesses a region of negative differential mobility for fields between 3×10^3 V cm^{-1} and about 5×10^4 V cm^{-1}. This arises as a result of electrons being scattered between two valleys in the GaAs conduction band, each having a different value of electron effective mass. The band structure for GaAs is illustrated in Fig. 2.4. For low fields the electrons are contained in

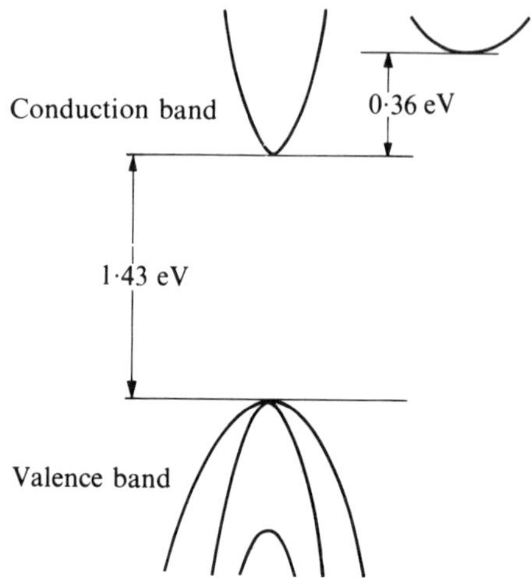

Fig. 2.4 Band structure of GaAs.

the lower valley where the effective mass is small and the mobility is large. As the field is increased the electron energy is increased, and electrons transfer to the upper valley where the effective mass is large and the mobility is small. As a result the average electron velocity decreases and a negative differential mobility is produced. We shall not consider this phenomenon further since it plays no significant role in the operation of the avalanche-diode oscillator. It is the basis of the Gunn or transferred-electron oscillator.

2.3 Impact ionization and avalanche multiplication

Theory of impact ionization

In most semiconductors of practical interest, fields in excess of 10^5 V cm^{-1} are required for the onset of avalanche multiplication. Even though the ionization rate is finite at much lower fields, it is very small, and an avalanche discharge cannot be maintained in any practical structure. Once an electron has reached the ionization threshold energy ϵ_i, it is still more likely to lose part of its total energ by the emission of an optical phonon and drop below threshold. For an electron with energy ϵ_i, the relative probability that the first collision is an ionizing collision is l_{op}/l_i, where l_i is the mean free path for producing ionization and l_{op} is the mean free path for optical-phonon emission. l_{op}/l_i for an electron at the threshold energy is typically 0·02 to 0·05.

The parameters used to describe impact ionization are the ionization rates α and β for electrons and holes respectively. The ionization rate is the average number of ionizing collisions experienced by a single electron (or hole) per unit distance of travel in the direction of the field. α and β are rapidly increasing functions of the electric field. In Ge [13, 14], Si [15], and GaAs [14], both α and β have been measured, and in all cases the experimental results can be fitted to to an expression of the form [16]

$$\alpha = A \exp\left(-\frac{b}{E}\right)^m, \tag{2.19}$$

where A and b are temperature dependent.

In Table 2.1 the room temperature values of A, b, and m are listed for both holes and electrons in the three semiconductors.

TABLE 2.1

Parameters describing ionization rates of electrons and holes in Ge, Si, and GaAs at 300°K.

Semiconductor	Carrier	A	b	m	l_{op}	ϵ_{op}
		cm^{-1}	$V\ cm^{-1}$		A°	eV
Ge	electron	$1{\cdot}55 \times 10^7$	$1{\cdot}56 \times 10^6$	1	65 ± 10	0·037
Ge	hole	$1{\cdot}0 \times 10^6$	$1{\cdot}28 \times 10^6$	1		
Si	electron	$3{\cdot}8 \times 10^6$	$1{\cdot}75 \times 10^6$	1	62 ± 5	0·063
Si	hole	$2{\cdot}25 \times 10^7$	$3{\cdot}26 \times 10^6$	1	45 ± 5	
GaAs	electron	$3{\cdot}5 \times 10^5$	$6{\cdot}85 \times 10^5$	2	35 ± 5	0·035
GaAs	hole	$3{\cdot}5 \times 10^5$	$6{\cdot}85 \times 10^5$	2		

This exponential form of the ionization rate was first discussed by Shockley [17] by considering a simple low-field model, assuming that the average electron energy is much smaller than the ionization energy. In this approximation the only electrons which produce ionizing collisions are those which are accelerated up to ϵ_i without suffering a single optical-phonon collision. Consider an electron accelerated from zero energy up to ϵ_i by the electric field E. It must travel through a path length $l = \epsilon_i/qE$. Since $\epsilon_i \gg \epsilon_{op}$, this electron has sufficient energy over almost its entire travel to emit an optical phonon, and the probability that it will reach the ionization threshold is therefore given by

$$P_0 = \exp\left(-\frac{l}{l_{op}}\right). \tag{2.20}$$

Having reached the threshold energy the electron is still more likely to have a phonon collision, and the probability for impact ionization is

$$P = \frac{l_{op}}{l_i} \exp\left(-\frac{l}{l_{op}}\right). \tag{2.21}$$

This is the probability that the electron will generate a hole–electron pair on one start from zero energy. Therefore the electron must make $1/P$ starts from zero energy to suffer one ionizing collision. The other $(P^{-1} - 1)$ starts result in the electron reaching the energy ϵ_{op}, emitting an optical phonon, and dropping back to zero energy. Thus the total energy supplied by the field to produce one ionizing collision is

$$\epsilon_T = \epsilon_i + \epsilon_{op}\left[\frac{l_i}{l_{op}} \exp\left(\frac{l}{l_{op}}\right) - 1\right]. \quad (2.22)$$

But, by the definition of the ionization rate, the total energy supplied by the field is also given by

$$\epsilon_T = \frac{qE}{\alpha}. \quad (2.23)$$

From eqns (2.23) and (2.24) a relationship for α is obtained, namely

$$\alpha = qE\left[\frac{1}{\epsilon_i + \epsilon_{op}\left\{\frac{l_i}{l_{op}} \exp\left(\frac{l}{l_{op}}\right) - 1\right\}}\right] \quad (2.24)$$

This expression has been derived using the approximation that the average electron energy is less than the optical-phonon energy $E < (\epsilon_{op}/ql_{op})$. Using this approximation in the above expression for α we get

$$\alpha = \left(\frac{qEl_{op}}{\epsilon_{op}l_i}\right) \exp\left(-\frac{\epsilon_i}{ql_{op}E}\right). \quad (2.25)$$

This low-field approximation is valid for fields up to the range 2×10^4 V cm^{-1} to 2×10^5 Vcm^{-1} for the three most commonly used semiconductors, Ge, Si, and GaAs, listed in Table 2.1.

At higher fields the theory can be extended to show [17] that $\ln \alpha$ is proportional to $1/E^2$. This analysis neglects phonon collisions for electron energies between ϵ_{op} and ϵ_i, and this is a drastic oversimplification. In addition, we have not considered the case where an electron gains an energy greater than $(\epsilon_i + \epsilon_{op})$, in which case a phonon collision followed by impact ionization is possible. Moll and Mayer [18] have considered these corrections to Shockley's theory and have indeed shown that they are not negligible. However, the above analysis illustrates the physics involved in the ionization process, and further elaboration of this theory does not concern us here.

Probably the most comprehensive theoretical treatment of impact ionization has been carried out by Baraff [19, 20] who obtained a general solution for the ionization rate in terms of the three parameters l_{op}, ϵ_i, and ϵ_{op} by solving the

Boltzmann transport equation. Even this detailed theoretical treatment involves several simplifying assumptions, including a constant mean free path and consideration of only optical-phonon and ionizing collisions. Baraff's results, which represent universal plots from which α can be obtained for any semiconductor when the above parameters are known, are presented in Fig. 2.5. Log $(\alpha\, l_{op})$ is shown as a function of ϵ_i/qEl_{op} for various values of the parameter ϵ_{op}/ϵ_i. The experimental results for Ge are shown in Fig. 2.6 along with the Baraff curve for $\epsilon_{op}/\epsilon_i = 0{\cdot}022$. The agreement is good. The mean free paths for optical-phonon

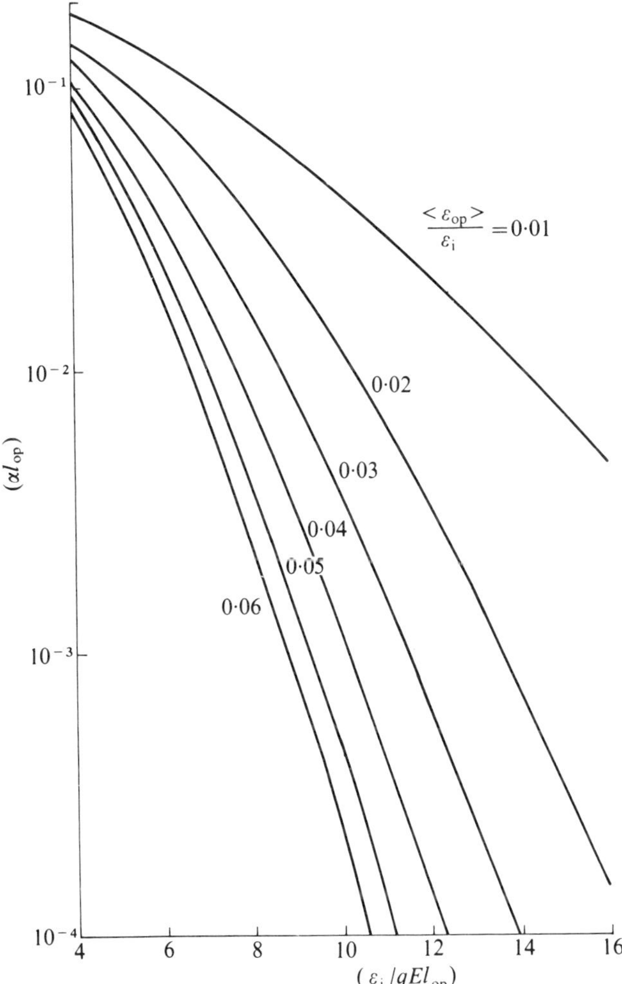

Fig. 2.5 Baraff plots of (αl_{op}) versus (ϵ_i/qEl_{op}) with $\langle\epsilon_{op}\rangle/\epsilon_i$ as a parameter (after Baraff [19]).

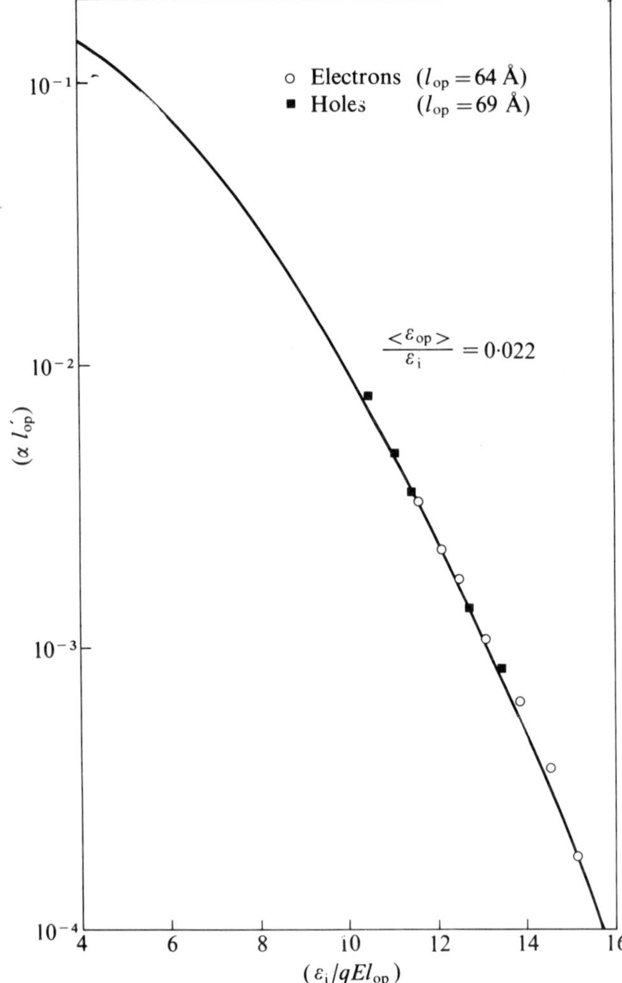

Fig. 2.6 Ionization rates for holes and electrons in Ge and the Baraff plot for $\langle\epsilon_{op}\rangle/\epsilon_i = 0{\cdot}022$ (after Logan and Sze [14]).

collisions listed in Table 2.1 are obtained by fitting the experimentally measured ionization rates to the appropriate Baraff curves for each of the semiconductors in turn. In Baraff's theory, optical-phonon absorption is excluded, it being assumed that in all electron–optical-phonon collisions the electron loses an energy ϵ_{op} by emitting an optical phonon. This assumption is certainly justified in those semiconductors, such as Si, in which ϵ_{op} is much larger than kT at room temperature. However, in Ge and GaAs, the thermal excitation of optical phonons at room temperature cannot be ignored and the absorption of optical phonons cannot be neglected. This modification to Baraff's theory has been considered by Crowell and Sze [21], who replaced the parameter ϵ_{op} with the average energy loss per optical-phonon collision $\langle\epsilon_{op}\rangle$, thereby allowing for the possibility of phonon absorption. $\langle\epsilon_{op}\rangle$ is given by

$$\langle\epsilon_{op}\rangle = \epsilon_{op} \tanh\left(\frac{\epsilon_{op}}{2kT}\right). \tag{2.26}$$

Baraff's curves are then modified by using $\langle\epsilon_{op}\rangle$ in place of ϵ_{op}.

In practice, avalanche-diode oscillators frequently work at temperatures well above the ambient where heating has a significant effect on the ionization process. The variation of α with temperature can be obtained from the modified Baraff theory, provided the temperature dependence of the three parameters $\langle\epsilon_{op}\rangle$, ϵ_i, and l_{op} is known. Since the band gap decreases with increasing temperature, as shown [12, 22] in Fig. 2.7 for Ge, Si, and GaAs, it follows that the ionization energy must also decrease, and this tends to increase the probability of an ionizing collision. On the other hand, the optical-phonon mean free path decreases as the temperature is raised, according to

$$l_{op} = l_o \tanh\left(\frac{\epsilon_{op}}{2kT}\right), \tag{2.27}$$

and this reduces α, since the electrons gain less energy between successive phonon collisions. The second effect dominates and since the optical-phonon energy is independent of temperature, α decreases with increasing temperature as shown in Fig. 2.8 for Si.

An approximate analytical expression has been derived by Crowell and Sze [21] to fit the Baraff curves, relating α to E and T, namely

$$\ln(\alpha l_{op}) = \begin{bmatrix} (11{\cdot}5r^2 - 1{\cdot}17r + 3{\cdot}9 \times 10^{-4})x^2 + \\ + (46r^2 - 11{\cdot}9r + 1{\cdot}75 \times 10^{-2})x + \\ + (-757r^2 + 75{\cdot}5r - 1{\cdot}92) \end{bmatrix} \tag{2.28}$$

where $r = \langle\epsilon_{op}\rangle/\epsilon_i$ and $x = \epsilon_i/qEl_{op}$. Provided $0{\cdot}01 < r < 0{\cdot}06$ and $5 < x < 16$ this expression agrees (within ±2 per cent) with the curves computed numerically by the Baraff theory.

Avalanche multiplication

The charge density generated by impact ionization in a semiconductor depends on the value of the electric field and the width of the high-field region. Let us consider a high-field region between $x = 0$ and $x = w$, in which avalanche multiplication is initiated by electrons entering at the cathode, $x = 0$, as shown in Fig. 2.9. This is essentially identical to the situation arising in the space-charge region at a p–n junction where the electrons are generated by thermal excitation in the neutral p-region and reach the cathode plane by diffusion. In Fig. 2.9 the electrons drift to the right, creating hole–electron pairs as they go, and the generated holes move to the left, producing additional hole–electron pairs by impact ionization. Thus, if the hole and electron ionization rates are

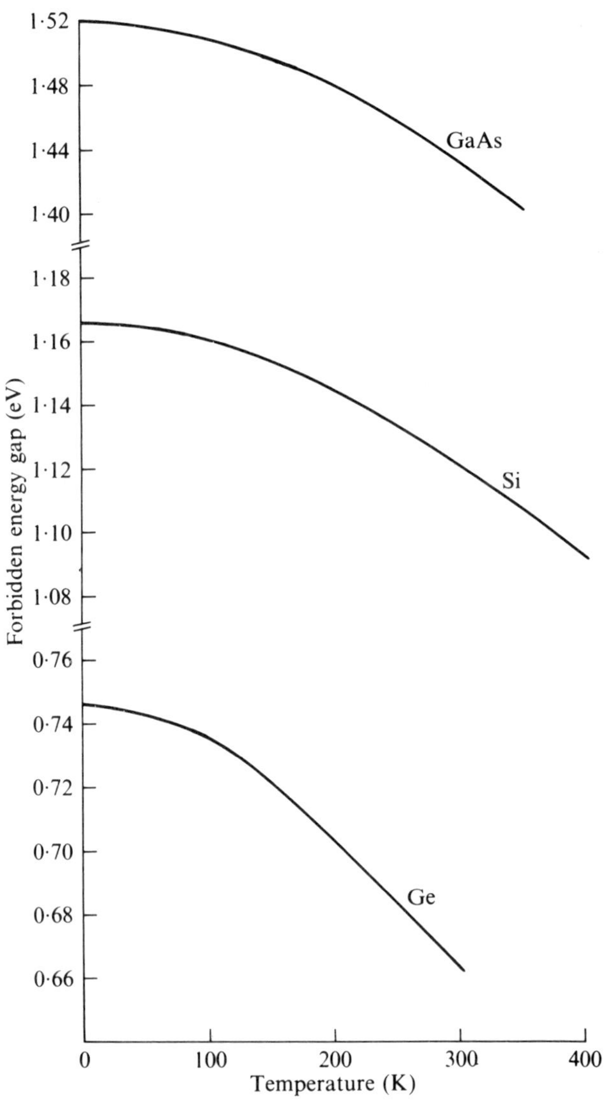

Fig. 2.7 Forbidden energy gap versus temperature for Ge, Si, and GaAs (after Smith [22] and Sze [12]).

approximately equal, the holes provide a feedback effect continuously replenishing the supply of electrons at the left-hand side of the region. The multiplication factor M_n for electrons is defined as the ratio of the electron density leaving the region at $x = w$ to the electron density entering at the cathode:

$$M_n = \frac{n(w)}{n(0)} = \frac{n}{n_0}. \tag{2.29}$$

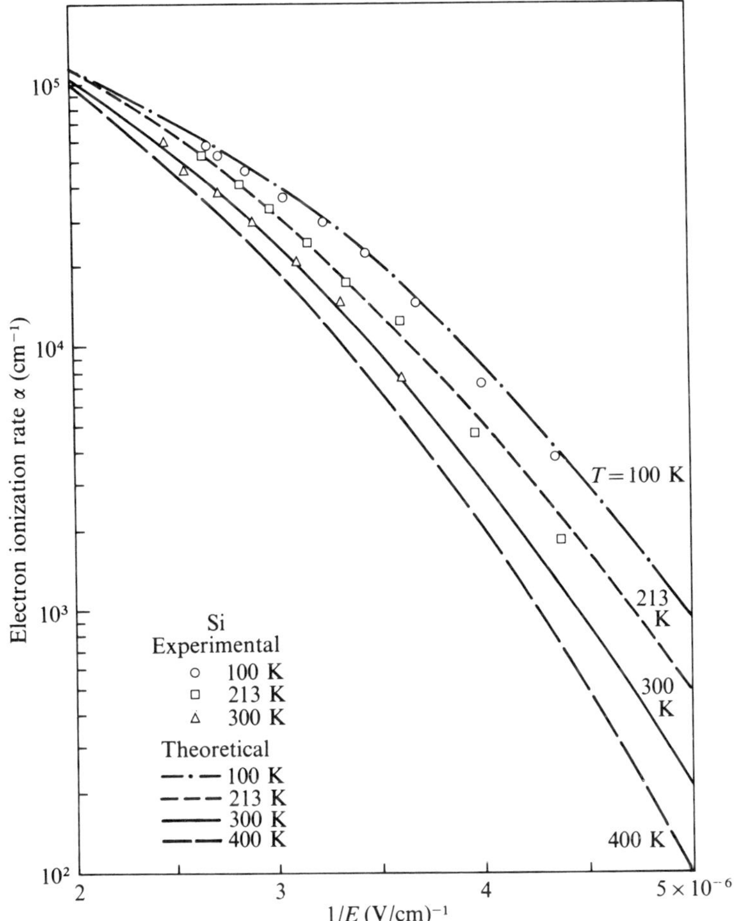

Fig. 2.8 Electron ionization rate versus reciprocal electric field for Si. (After Crowell and Sze [21]).

Consider a small element of width dx within the region with an electron density $n_1(x)$ entering from the left, and a hole density $p_1(x)$ entering from the right, as illustrated in Fig. 2.9. The total number of electrons generated within dx, by both the incident holes and electrons, is

$$\mathrm{d}n_1 = \alpha n_1\,\mathrm{d}x + \beta p_1\,\mathrm{d}x. \tag{2.30}$$

The total number of electrons collected at the anode $x = w$ is $(n_1 + p_1)$, since dx is assumed to be small and holes and electrons are always generated in pairs.

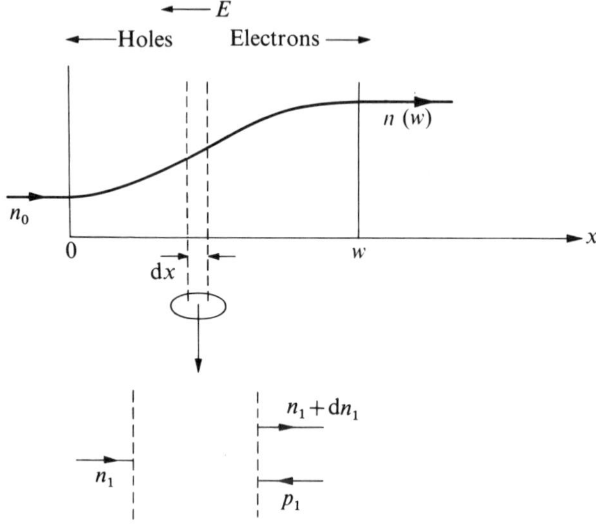

Fig. 2.9 Charge multiplication in high-field region of a semiconductor.

Thus, eqn (2.30) can be rewritten as

$$\frac{dn_1}{dx} = (\alpha - \beta)n_1 + \beta n, \tag{2.31}$$

where $n = n_1 + p_1$. An expression relating M_n to α and β can now be derived by integration of eqn (2.31) subject to the boundary conditions

$$\begin{aligned} n_1 &= n_0 \text{ at } x = 0, \\ n_1 &= n = M_n n_0 \text{ at } x = w. \end{aligned} \tag{2.32}$$

Equation (2.31) has the general form

$$n_1' + Pn_1 = Q, \tag{2.33}$$

which has the solution

$$n_1 = \frac{\int_0^x Q\left[\exp\int_0^x P\,dx'\right]dx + \text{constant}}{\exp\int_0^x P\,dx}. \tag{2.34}$$

Therefore n_1 is given by

$$n_1 = \frac{\int_0^x \beta n\left[\exp\int_0^x (\beta-\alpha)\,dx'\right]dx + \text{constant}}{\exp\int_0^x (\beta-\alpha)\,dx}, \tag{2.35}$$

and using (2.32) we obtain an expression for M_n, namely

$$M_n^{-1} = \exp \int_0^w (\beta - \alpha)\,dx - \int_0^w \beta \left[\exp \int_0^x (\beta - \alpha)\,dx'\right] dx. \qquad (2.36)$$

In order to reduce this expression to the more normal form we must use the change of variable

$$y = \int_0^x (\beta - \alpha)\,dx$$

to show that

$$\int_0^w (\beta - \alpha)\left[\exp \int_0^x (\beta - \alpha)\,dx'\right] dx = \exp\left[\int_0^w (\beta - \alpha)\,dx\right] - 1. \qquad (2.37)$$

Using (2.37) to substitute for the first term on the right-hand side of (2.36), we obtain the commonly-used form for the multiplication factor

$$1 - \frac{1}{M_n} = \int_0^w \alpha \left[\exp \int_0^x (\beta - \alpha)\,dx'\right] dx. \qquad (2.38)$$

A similar expression can be obtained for the hole multiplication factor M_p by considering the avalanche to be initiated by holes entering the region from the anode. The result is

$$1 - \frac{1}{M_p} = \int_0^w \beta \left[\exp \int_x^w (\alpha - \beta)\,dx'\right] dx. \qquad (2.39)$$

We shall use eqns (2.38) and (2.39) in the next chapter to derive the condition for avalanche breakdown in a junction.

References

1. Shockley, W. *Bell Syst. tech. J.* **30**, 990 (1951).
2. Moll, J. L. *Physics of semiconductors.* McGraw-Hill, New York (1964).
3. Conwell, E. and Weisskopf, V. F. Theory of impurity scattering in semiconductors. *Phys. Rev.* **77**, 388 (1950).
4. Sze, S. M. and Irvin, J. C. Resistivity, mobility and impurity levels in GaAs, Ge and Si at 300°K, *Solid-St. Electron.* **11**, 599 (1968).
5. Seidel, T. E. and Scharfetter, D. L. Dependence of hole velocity upon electric field and hole density for p-type silicon, *J. Phys. Chem. Solids* **28**, 2563 (1967).
6. Norris, C. B. and Gibbons, J. F. Measurement of high-field carrier drift velocities in Si by a time-of-flight technique, *IEEE Trans. Electron Devices* **ED–14** 38 (1967).
7. Duh, C. Y. and Moll, J. L. Electron drift velocity in avalanching silicon diodes. *IEEE Trans. Electron Devices* **ED-14**, 46 (1967).

8. Duh, C. Y. and Moll, J. L. Temperature dependence of hot electron drift velocity in silicon at high electric field. *Solid-St. Electron.* **11**, 917 (1968).
9. Ruch, J. G. and Kino, G. S. Measurement of the velocity–field characteristic of gallium arsenide *Appl. Phys. Lett*, **10**, 40 (1967).
10. Dalal, V. L. Hole-velocity in p-GaAs, *Appl. Phys. Lett*, **16**, 489 (1970).
11. Canali, C., Ottaviani, G. and Quaranta, A. A. Drift velocity of electrons and holes and associated anisotropic effects in silicon, *J. Phys. Chem. Solids* **32**, 1707 (1971).
12. Sze, S. M. *Physics of semiconductor devices*, Wiley (1969).
13. Miller, S. L. Avalanche breakdown in Germanium, *Phys. Rev.* **99**, 1234 (1955).
14. Logan, R. A. and Sze, S. M. Avalanche multiplication in Ge and GaAs p–n junctions, *J. Phys. Soc. Japan Supplement.* **21**, 434 (1966).
15. Lee, C. A., Logan, R. A., Batdorf, R. L., Kleimack, J. J. and Wiegmann, W. Ionization rates of holes and electrons in silicon. *Phys. Rev.* **134**, 761 (1964).
16. Sze, S. M. and Gibbons, G. Avalanche breakdown voltages of abrupt and linearly graded p–n junctions in Ge, Si, GaAs, and GaP. *Appl. Phys. Lett.* **8**, 111 (1966).
17. Shockley, W. Problems related to p–n junctions in silicon. *Solid-St. Electron.* **2**, 35 (1961).
18. Moll, J. L. and Mayer, N. I. *Solid-St. Electron.* **3**, 155 (1961).
19. Baraff, G. A. Distribution functions and ionization rates for hot electrons in semiconductors. *Phys. Rev.* **128**, 2507 (1962).
20. Baraff, G. A. Maximum anisotropy approximation for calculating electron distributions; application to high field transport in semiconductors. *Phys. Rev.* **133**, A26 (1964).
21. Crowell, C. R. and Sze, S. M. Temperature dependence of avalanche-multiplication in semiconductors. *Appl. Phys. Lett.* **9**, 242 (1966).
22. Smith, R. A. *Semiconductors.* Cambridge University Press (1959).

3 Reverse-biased p–n junction

3.1 Voltage-current characteristic

When a reverse-bias voltage is applied to a p–n junction the voltage is dropped, almost entirely, across a narrow region at the junction called the space-charge layer. The electric field has a maximum value at the metallurgical junction and decreases on either side with a gradient determined by the doping concentration in the semiconductor. When the reverse voltage is increased, majority carriers are removed from the edges of the space-charge layer and the layer widens, causing a displacement current to flow, determined by the rate of increase of the electric field. The field distribution in the space-charge layer of a p^+n junction for two values of reverse-bias voltage is shown in Fig. 3.1. At any value of bias the applied voltage is simply the area under the field curve. In the case of the p^+n junction the extension of the space-charge layer into the p^+-region is small by comparison with the spreading into the much lower-doped n-region. This type of junction is referred to as a one-sided abrupt junction.

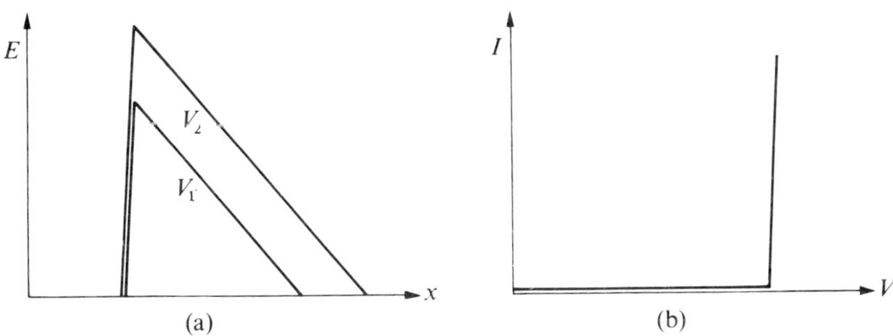

Fig. 3.1 (a) Electric field E versus distance x, in the space-charge region of an abrupt junction for two values of reverse-bias voltage. (b) Reverse-bias current I versus V for a p–n junction.

Also shown in Fig. 3.1 is the current–voltage characteristic of the diode under reverse bias. A small constant current flows for voltages below the avalanche breakdown voltage V_B. In an ideal diode this current is produced by minority carriers, thermally generated in the neutral n and p^+ regions, which flow by diffusion toward the edges of the space-charge layer. These carriers then drift across the space-charge layer into the opposite neutral region where they are majority carriers and produce a majority-carrier current. As holes and electrons are extracted from the regions in which they are minority carriers, the minority-carrier concentrations at the edges of the space-charge layer are depressed below their equilibrium values as shown in Fig. 3.2. The regions of

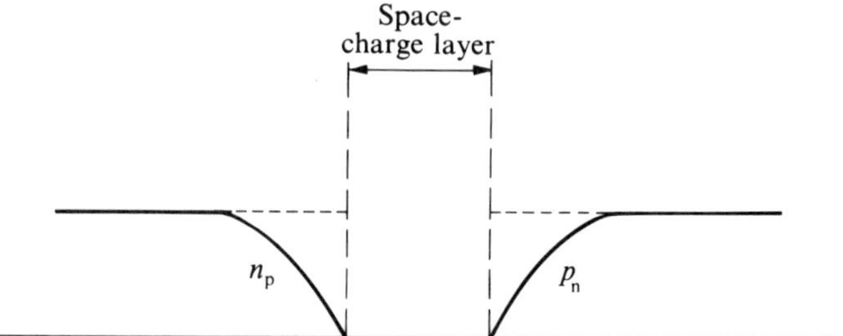

Fig. 3.2 Minority carrier distributions in neutral regions close to a p–n junction.

minority-carrier depression extend into the neutral regions about one diffusion length from the edges of the space-charge layer. The minority-carrier concentration gradients give rise to the diffusion current, and the majority carriers generated in this region flow away from the junction, constituting a majority-carrier current. The electron and hole currents in the diode are illustrated in Fig. 3.3. The total current is of course independent of position. A small reverse

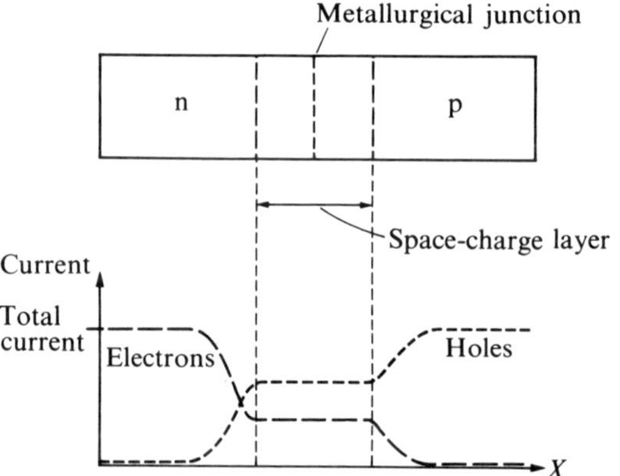

Fig. 3.3 Electron, hole, and total currents in reverse-biased p–n junction.

voltage is sufficient to reduce the minority-carrier concentrations at the edges of the space-charge layer to a very low level, and increasing the reverse voltage produces little change in the carrier concentrations and little change in the current. When the voltage is increased to the breakdown value, a rapid increase in current is produced by the onset of avalanche multiplication. The charge multiplication is initiated by the thermally generated carriers which comprise the saturation current. The saturation current density J_s for a p⁺n diode is inversely proportional to the donor concentration N_D, and is given bv

$$J_s = \sqrt{\left(\frac{D_p}{\tau_p}\right)} \frac{n_i^2}{N_D},$$

where D_p is the hole diffusion coefficient, τ_p is the lifetime for hole recombination, in the n-region, and n_i is the intrinisic carrier concentration.

In addition to the thermal diffusion current, the current flow at voltages below breakdown, in many practical cases, is also determined by (a) carrier generation within the active region, (b) tunnelling between states in the band gap, and (c) surface leakage currents. When the thermal diffusion current is dominant, the current–voltage characteristic is given by the ideal diode equation given by Shockley [1], namely

$$J = J_s\left(\exp\frac{qV}{kT} - 1\right). \tag{3.1}$$

This equation accurately describes the I–V relationship for Ge diodes. However, for Si and GaAs devices, (a) and (b) play a major role. The current produced by carrier generation in the space-charge layer is given by

$$J_{gen} = \int_0^w qg \, dx, \tag{3.2}$$

where $g = n_i/\tau_e$ is the carrier generation rate given in terms of the intrinsic carrier concentration and an effective lifetime for recombination τ_e. By integrating over the width of the space-charge layer w, the generated current is

$$J_{gen} = \frac{qn_iw}{\tau_e}, \tag{3.3}$$

which is proportional to the space-charge layer width and therefore depends on the reverse-bias voltage. The total reverse current for a p⁺n junction is then

$$J = J_s + J_{gen}$$

$$= q\sqrt{\left(\frac{D_p}{\tau_p}\right)} \frac{n_i^2}{N_D} + \frac{qn_iw}{\tau_e}. \tag{3.4}$$

For semiconductors with a large intrinsic carrier concentration, the thermal diffusion current is much larger than the current generated in the active region and the reverse characteristic follows Shockley's equation. If n_i is small, as for example in Si, the generation term dominates at room temperature.

In many practical devices excess currents flow by surface leakage effects at the edge of the diode. These can result from ionic conduction due to contamina-

tion on the surface of the diode, by surface charge layers induced by charges external to the semiconductor, or by the presence of high-field regions at the edges of the diode. Careful preparation of the devices and shaping of the diode to reduce the surface fields can, in most cases, reduce the surface leakage currents to tolerable levels.

Beyond breakdown the current–voltage characteristic of the junction is determined by the resistance of the space-charge layer, the parasitic resistance of the neutral regions on either side of the active region, and the effects of internal heating on the breakdown voltage of the junction. These effects are discussed in §3.5.

3.2 The depletion-layer approximation

The field and potential distribution in the space-charge layer of a p–n junction can be calculated analytically using the so-called depletion-layer approximation in which it is assumed that the mobile charge is fully depleted from the layer. The depletion approximation is illustrated in Fig. 3.4, which shows the impurity

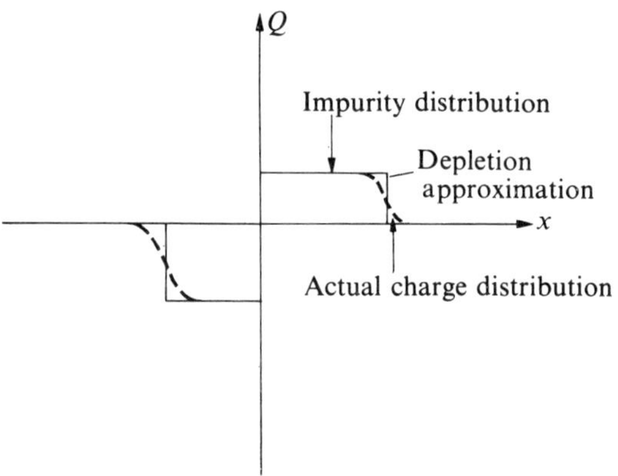

Fig. 3.4 Depletion approximation: impurity distribution and actual charge distribution in step junction.

distribution and resultant charge distribution close to the metallurgical junction. In the regions adjacent to the junction the actual charge density agrees exactly with the depletion approximation and is equal to the net ionized impurity density. However, near the edges of the space-charge layer the actual distribution of charge deviates from the ionized impurity density by the presence of the mobile charge, whereas in the depletion approximation the charge density is assumed to be exactly equal to the ionized impurity density right up to the edge of the layer. Under reverse bias the depletion approximation yields results which agree well with the experimentally measured properties of the space-charge layer.

The total voltage across the diode depletion layer is the sum of the built-in voltage V_b and the applied voltage V_A, where

$$V_b = \frac{kT}{q} \ln\left(\frac{N_A N_D}{n_i^2}\right). \tag{3.5}$$

The high-field region is confined to the depletion layer, and the total negative charge on the ionized acceptors on the p-side of the junction is exactly balanced by the total positive charge on the ionized donors on the n-side.

$$|N_A w_1| = |N_D w_2|. \tag{3.6}$$

The electric field and potential distribution in the depletion region are obtained from Poisson's equation

$$\frac{dE}{dx} = \frac{\rho}{\epsilon}, \qquad E = -\frac{dV}{dx}, \tag{3.7}$$

with the boundary conditions $E = 0$ at $x = -w_1$ and $x = w_2$. On the p-side of the junction the electric field $E_1(x)$ is given by integrating

$$\frac{dE_1}{dx} = -\frac{qN_A}{\epsilon}, \tag{3.8}$$

and on the n-side the field $E_2(x)$ is given by integrating

$$\frac{dE_2}{dx} = \frac{qN_D}{\epsilon}, \tag{3.9}$$

with the result

$$E_1(x) = -\frac{qN_A}{\epsilon}(x + w_1) \tag{3.10}$$

and

$$E_2(x) = \frac{qN_D}{\epsilon}(x - w_2). \tag{3.11}$$

The maximum field, at the metallurgical junction $x = 0$, is given by

$$E_M = -\frac{qN_A w_1}{\epsilon}. \tag{3.12}$$

The field is negative because of our choice of diode polarity with respect to the positive direction of the x-axis. The potential distribution is obtained by integrating the field. We consider the left-hand edge of the space-charge layer to be at zero potential. Then the potential on the p-side is

$$V_1(x) = \frac{qN_A}{\epsilon}\left[\frac{x^2}{2} + \frac{w_1^2}{2} + w_1 x\right], \tag{3.13}$$

and on the n-side is

$$V_2(x) = \frac{qN_A w_1^2}{2\epsilon} - \frac{qN_D}{\epsilon}\left[\frac{x^2}{2} - w_2 x\right]. \qquad (3.14)$$

The potential distribution in the depletion layer is illustrated in Fig. 3.5. From eqns (3.13) and (3.14) the total potential drop across the depletion layer is

$$V = \frac{qN_A w_1^2}{2\epsilon}\left[1 + \frac{N_A}{N_D}\right], \qquad (3.15)$$

or

$$V = \frac{qN_A w^2}{2\epsilon} \bigg/ \left[1 + \frac{N_A}{N_D}\right]. \qquad (3.16)$$

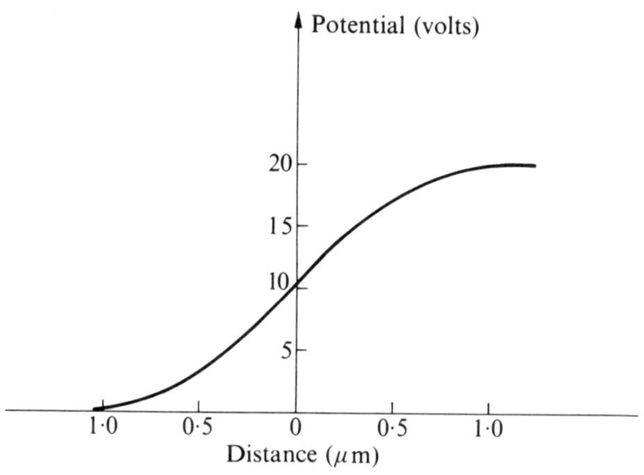

Fig. 3.5. Potential distribution in Si symmetrical step junction ($|N_A| = |N_D| = 10^{16}$ cm^{-3}) for a reverse-bias voltage of 20 volts.

The capacitance associated with the depletion region is defined as $C = dQ/dV$, where Q is the total space charge on one side of the junction ($Q = qAN_A w_1$). Using eqn (3.15) it is easily shown that $C = \epsilon A/w$, which is just the capacitance of two parallel plates located at the edges of the depletion layer.

3.3 The avalanche breakdown condition

Avalanche breakdown occurs in the junction when the electric field is large enough that the charge multiplication factor becomes infinite. Using the expressions derived in §2.3. relating the charge multiplication factor to the ionization rates, we arrive at the avalanche breakdown condition ($M_n, M_p \to \infty$), namely

$$\int_0^w \beta \exp\left[-\int_0^x (\beta - \alpha)\,\mathrm{d}x'\right]\mathrm{d}x = 1 \tag{3.17}$$

and

$$\int_0^w \alpha \exp\left[-\int_0^w (\alpha - \beta)\,\mathrm{d}x'\right]\mathrm{d}x = 1. \tag{3.18}$$

Both conditions are satisfied simultaneously. When the ionization rates for electrons and holes are equal ($\alpha = \beta$), the condition for breakdown reduces to the simple form

$$\int_0^w \alpha\,\mathrm{d}x = 1. \tag{3.19}$$

Solving the breakdown condition (3.17) or (3.18), allows the avalanche breakdown voltage to be calculated for any diode in which the field distribution in the active region is known. For most practical diodes the integration must be carried out by numerical methods. An exception is the p–i–n diode, shown in Fig. 3.6, in which the field is uniform throughout the depletion region extending

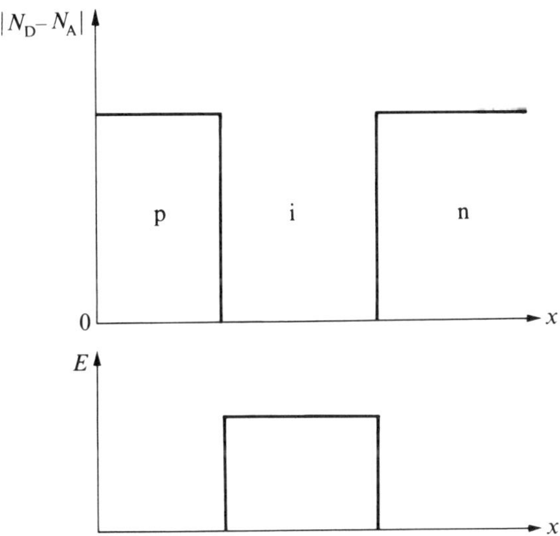

Fig. 3.6 Doping density versus distance and electric field profile under reverse bias for a p–i–n diode.

from the p–i interface to the n–i interface. The charge multiplication takes place uniformly throughout the active region, and if the ionization rates are equal the breakdown condition for a diode of width w is

$$\alpha w = 1. \tag{3.20}$$

Provided α is known as a function of the electric field the breakdown field and voltage are determined.

3.4 Avalanche breakdown of IMPATT and TRAPATT diodes

One-sided abrupt junctions

The structure most commonly used for practical IMPATT oscillators is the simple one-sided abrupt-junction diode discussed in §3.2. The breakdown voltage for this type of diode is plotted in Fig. 3.7 for Ge, Si, GaAs, and GaP [2]. The corresponding depletion-layer widths and maximum electric field values at breakdown are shown in Fig. 3.8. These results were obtained by numerical integration of the breakdown condition using the field distribution of Fig. 3.1 along with the ionization rates given in Table 2.1. The breakdown field is higher for the larger band-gap materials, reflecting the larger energy required to produce impact ionization. The breakdown voltage decreases as the

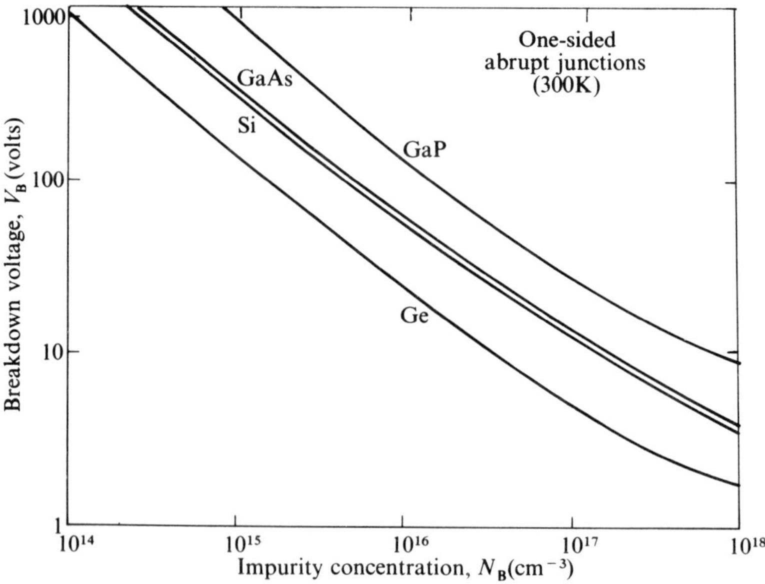

Fig. 3.7 Avalanche breakdown voltage versus impurity concentration for one-sided abrupt junctions in Ge, Si, GaAs, and GaP. (After Sze and Gibbons [2]).

doping density N increases, proportional to $(1/N)^{0.7}$. In contrast, the breakdown field changes by only a factor of two for a three-order-of magnitude change in the doping density. This reflects the rapid increase of α with field, requiring only a small increase in field to obtain breakdown in the narrower, more heavily doped structures. In the one-sided abrupt junction, charge multiplication does not take place uniformly throughout the space-charge layer but is concentrated in the region close to the metallurgical junction where the field is highest. The ionization rate, as a function of position in a typical p^+–n GaAs diode at breakdown, is shown in Fig. 3.9; α decreases rapidly, away from the junction, falling by a factor of approximately 100 in a distance of about 30 per cent of the total space-charge layer width. At breakdown, the field throughout the structure

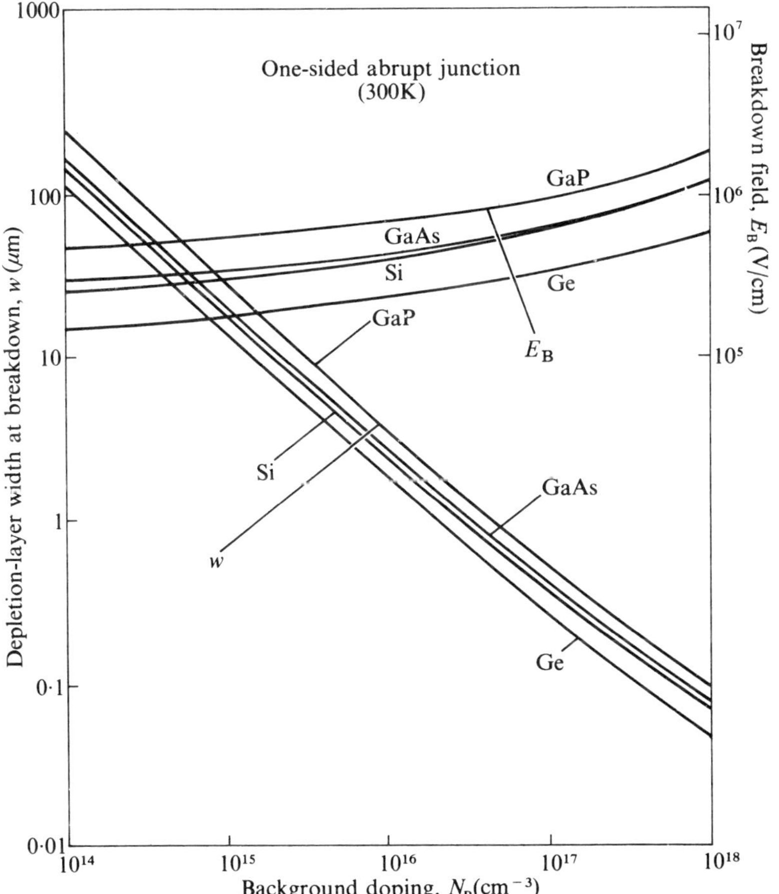

Fig. 3.8 Depletion-layer width and maximum field at breakdown for one-sided abrupt junctions in Ge, Si, GaAs, and GaP. [2].

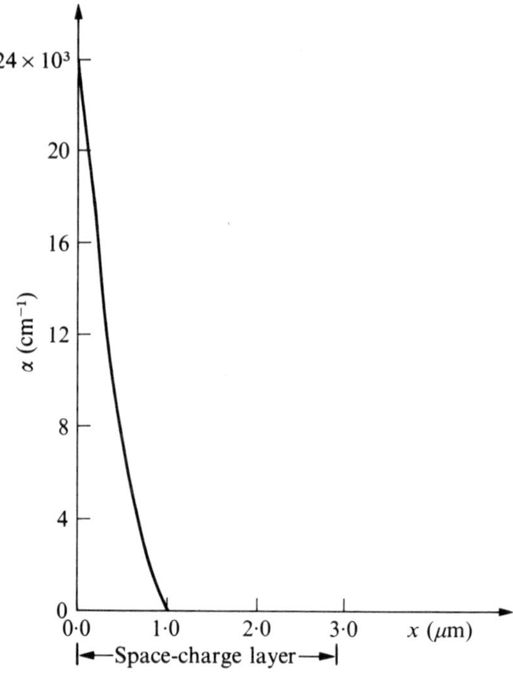

Fig. 3.9. Ionization rate α versus distance from the metallurgical junction x in a GaAs p^+–n diode ($N_D = 10^{16}$ cm^{-3}) at breakdown.

is sufficiently high that the carriers are drifting with the saturated drift velocity everywhere except for small regions at each end of the space-charge layer as shown in Fig. 3.10. Thus we can divide the active region of this diode into two zones in series, one occupying about 30 per cent of the total layer width, in which the electrons and holes are generated by impact ionization, and one which is free of carrier generation but in which the carriers drift with constant velocity. These regions are commonly referred to as the avalanche zone and drift zone respectively. This model is used in Chapter 4 where the negative-resistance properties of particular avalanche-diode structures are shown to depend on the ratio of the avalanche-zone width to the drift-zone width for each structure.

Avalanche breakdown cannot take place at voltages less than ϵ_i/q, since holes and electrons cannot then gain sufficient energy from the field to produce an ionizing collision. In heavily doped narrow structures, extremely high fields ($\sim 10^6$ V cm^{-1}) are reached before this voltage is exceeded. At these values of field, direct band-to-band transitions produced by field emission of electrons from the valence band become appreciable, and the junction breaks down by this process, called Zener tunnelling. Although there are no available states in the forbidden energy gap, each electron in the valence band has a small but finite

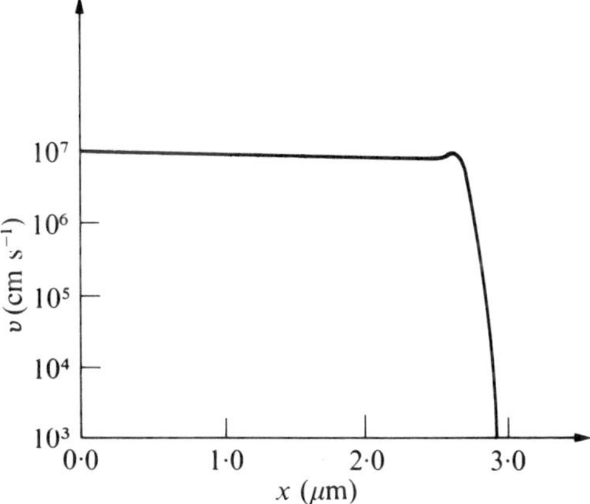

Fig. 3.10. Electron drift velocity v versus distance from the metallurgical junction x in the space-charge region of a GaAs p^+–n diode ($N_D = 10^{16}$ cm^{-3}) at breakdown.

probability of tunnelling into the conduction band. The quantum mechanical tunnelling of an electron through the forbidden gap is essentially identical to a particle tunnelling through a barrier, and can be analysed by the WKB approximation. The exact form of the potential barrier for an electron in the forbidden gap is not known in detail, but the tunnelling probability is not particularly sensitive to the shape of the barrier. For a triangular barrier the tunnelling probability is

$$P = \exp\left(-\frac{8\pi\sqrt{(2m^*)}E_g^{\frac{3}{2}}}{3qhE}\right),$$

where E is the electric field. For voltages less than $4\,E_g/q$ the junction breaks down by Zener tunnelling. For voltages greater than $6E_g/q$ the junction breaks down by the charge multiplication process. In the transition region between these values, both field emission and impact ionization contribute to the junction breakdown.

Symmetrical step junctions

At high frequencies, a symmetrical step junction is often used (for improved microwave performance) in preference to the one-sided abrupt junction discussed above. The improved performance of this structure is discussed in some detail in Chapter 5. In this chapter we are concerned with the breakdown characteristics of this type of diode, called a double-drift diode, shown in Fig. 3.11. The space-charge layer extends equally on both sides of the metallurgical junction creating an avalanche zone in the centre of the depletion layer with a drift zone on either side. In the structure shown in Fig. 3.11, the holes

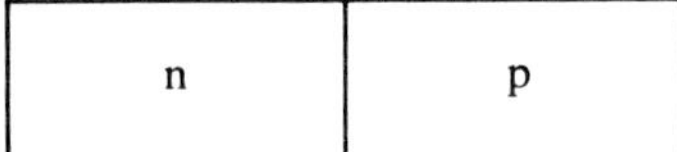

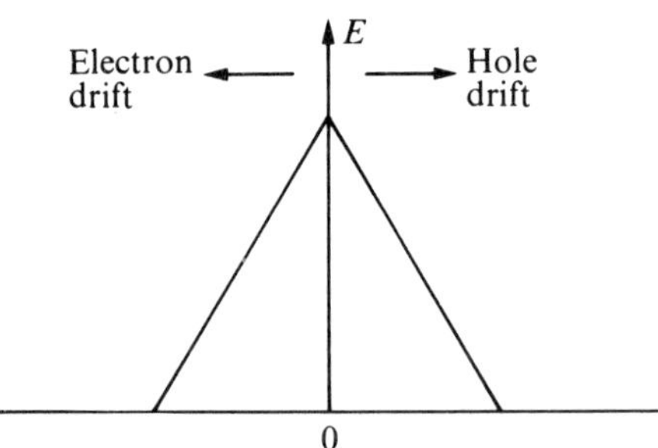

Fig. 3.11. Electric field profile in symmetrical ($N_A = N_D$) step junction (double-drift diode).

generated in the avalanche zone drift to the right, the electrons to the left. The breakdown voltages for symmetric step junctions are shown in Fig. 3.12 for Si and GaAs [3]. The depletion-layer widths and maximum fields at breakdown are shown for this structure in Fig. 3.13. It should be noted that the breakdown voltage for the double-drift diode is not equal to twice the value for the single-drift diode (Fig. 3.7) as might be expected at first. Furthermore, the breakdown field in the double-drift diode is approximately ten per cent smaller than the field at breakdown in the single-drift diode with the same value of doping density. This can be understood by examining the integrand in the breakdown condition. To compensate for the wider space-charge layer of the double-drift diode, a lower value of α and therefore a lower value of field will produce breakdown. Fig. 3.14 shows the ionization rate throughout the active region of a typical GaAs double-drift diode. The area under the $\alpha(x)$ curve is equal to unity at breakdown. Equal contributions are made to the integral from both sides of the metallurgical junction.

The Read diode

The diode structure proposed originally by Read [4] was a p^+–n–i–n^+ configuration in which the field distribution at breakdown is specifically designed to produce a highly localized avalanche zone in series with a constant-field drift zone as shown in Fig. 3.15. In practice, however, the Read diode is difficult to construct and although Ge and Si Read structures were fabricated several years ago, no real interest has been generated in these devices from a practical

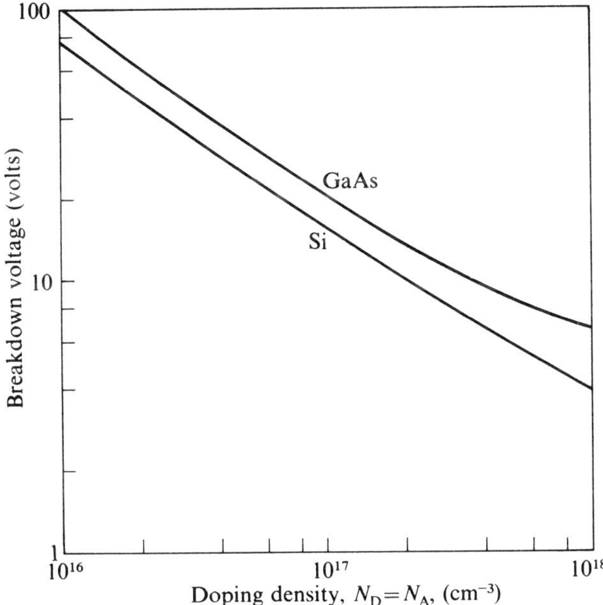

Fig. 3.12 Breakdown voltage versus doping density for Si and GaAs symmetrical step junctions (after Geraghty [3]).

or commercial point of view. The main difficulty in constructing this diode is the precise control required of the doping density and thickness of the n-region. If this region is too thin, or too lightly doped, the field in the intrinsic region is high enough, when the diode is biased to breakdown, to allow carrier generation to take place in the intrinsic region. On the other hand, if the n-region is too thick or too heavily doped, the p^+–n junction will break down before the depletion-layer edge penetrates into the intrinsic region. These limits are illustrated in Figs. 3.16 and 3.17, which show the breakdown voltages of Ge and Si Read diodes as a function of the thickness and doping level in the n-region, [5]. For example, for a Si Read structure with 10^{16} atoms/cm^3 in the n-region, the thickness of this region must be in the range of 1 μm to 2·8 μm. Otherwise the diode will behave like a p^+–n or a p^+–i–n^+ diode as illustrated in the figure.

TRAPATT diodes

For TRAPATT oscillators the structure most commonly used is the p^+–n–n^+ punch-through diode shown in Fig. 3.18, in which the depletion layer sweeps across the n-layer and into the n^+region before the field at the metallurgical junction reaches the breakdown value. For this structure, we define a punch-through factor $F = w_B/w$ [6] as the ratio of the depletion-layer width of the

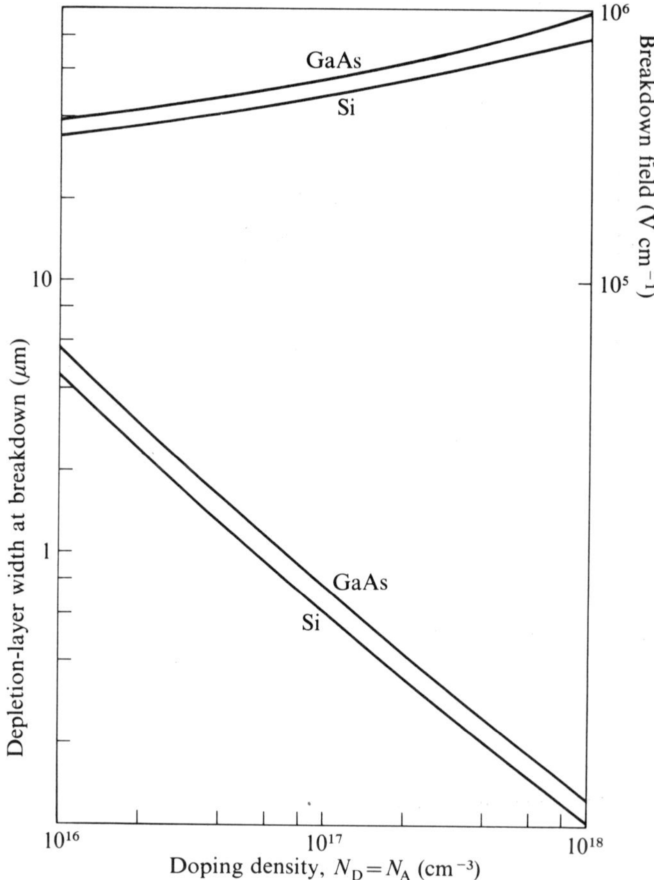

Fig. 3.13 Breakdown field and depletion-layer width at breakdown versus doping density for Si and GaAs symmetrical step junctions (after Geraghty [3]).

one-sided abrupt-junction diode to the width of the n-region in the punch-through diode with the same doping concentration. We have already seen that the dominant contribution to the avalanche multiplication in the one-sided abrupt junction originates in the region adjacent to the junction, of width x_A, occupying approximately one-third of the total depletion layer. As a result, for punch-through factors of approximately three or less, the avalanche region in the punch-through diode is identical to that in the one-sided abrupt junction with the same doping. However, as the width of the n-region is further reduced, a larger value of ionization rate is necessary to maintain avalanche breakdown,

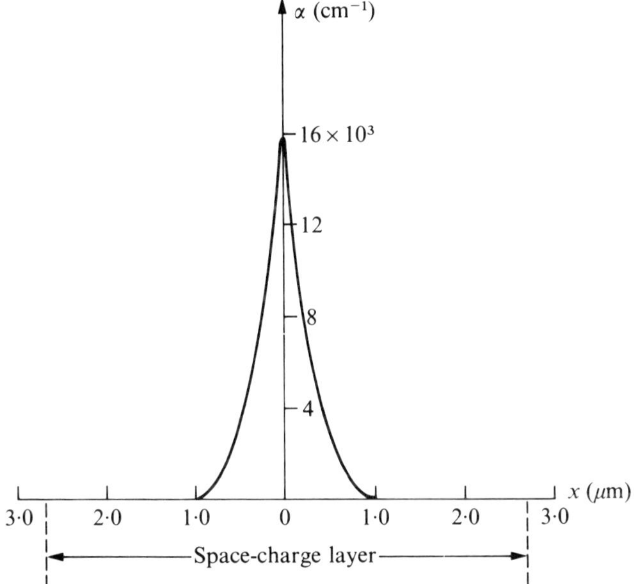

Fig. 3.14 Ionization rate α versus distance x in the space-charge layer of a double-drift GaAs diode ($|N_A| = |N_D| = 10^{16}$ cm^{-3})

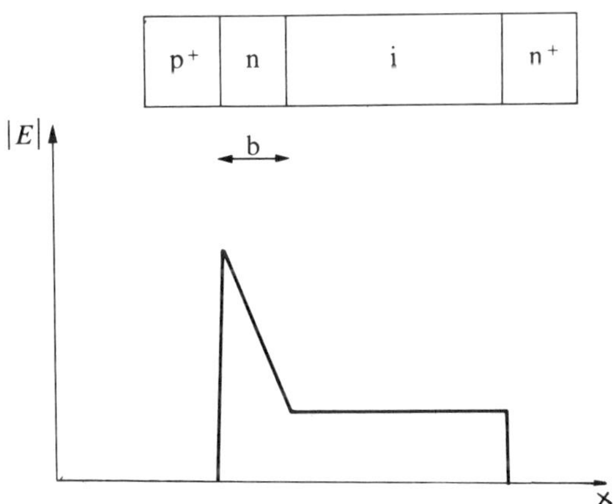

Fig. 3.15 Electric field profile in space-charge region of Read diode at breakdown (after Read [4]).

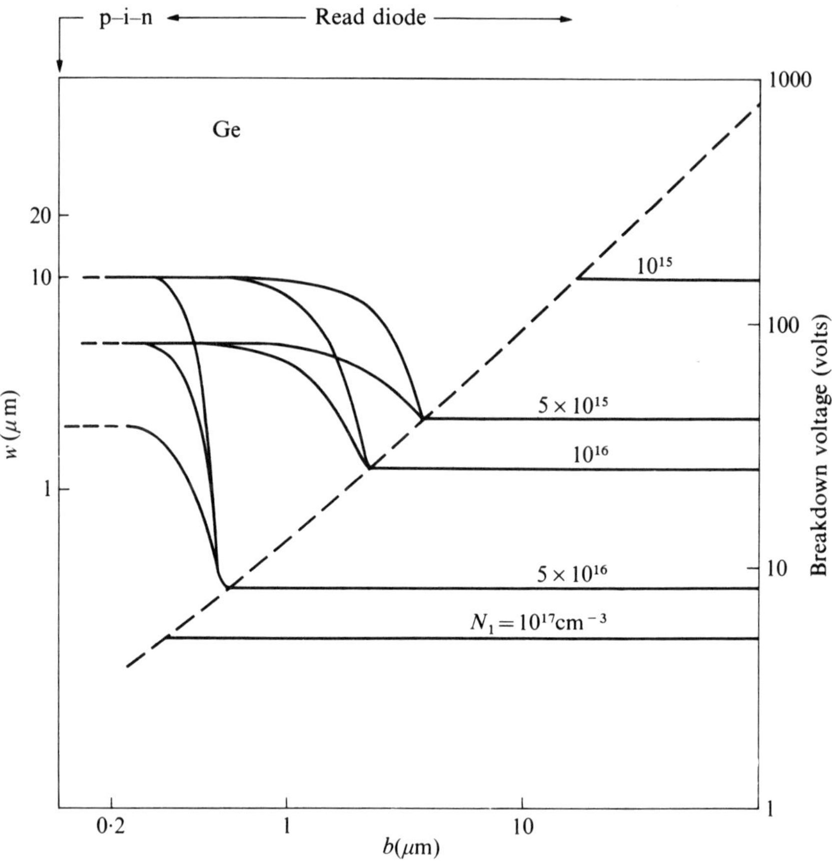

Fig. 3.16 Breakdown voltage versus b (the width of the n-region) for Ge Read and p–i–n diodes. The total depletion width, w and the doping level in the n-region, N_1, are parameters. (After Gibbons and Sze, [5]).

and the breakdown field is increased. The breakdown voltage for the punch-through diode, obtained by calculating the area under the field curve in Fig. 3.18 is given by

$$V_B = wE_B - \frac{q}{2\epsilon} Nw^2, \tag{3.21}$$

where E_B is the breakdown field for the one-sided abrupt junction with doping density N, provided $w > x_A$. For narrower structures, where $w < x_A$, the increased breakdown field must be calculated numerically since eqn (3.21) as it stands slightly underestimates the breakdown voltage.

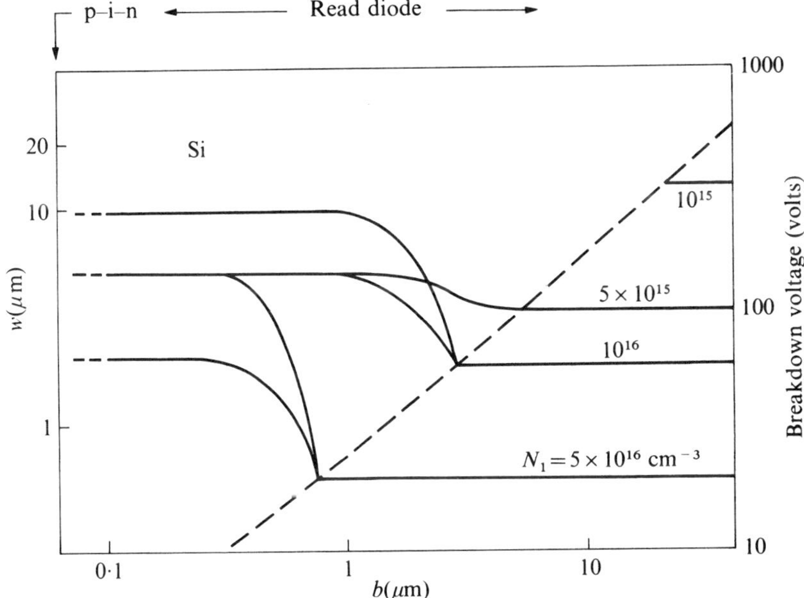

Fig. 3.17 Breakdown voltage versus b for Si Read and p–i–n diodes. The total depletion width, w, and the doping level in the n-region, N_1, are parameters [5]).

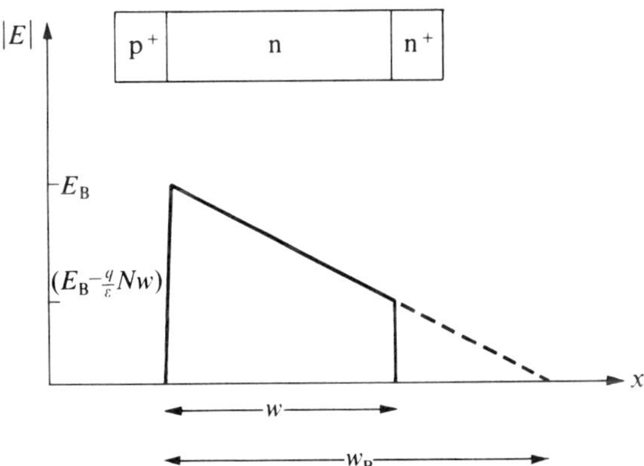

Fig. 3.18 Electric field profile in space-charge region of punch-through TRAPATT diode.

3.5 Space-charge resistance and temperature effects

The static breakdown voltage, discussed above for various types of junctions, is defined as the voltage for infinite charge multiplication. This assumption precludes any knowledge of the detailed shape of the current–voltage characteristic beyond breakdown. The d.c. current flowing in the junction at breakdown is in fact limited by three effects, namely

(a) The parasitic resistance R_p, associated with the neutral regions and the contacts,

(b) The space-charge resistance R_{SC}, associated with the flow of carriers through the space charge region, and

(c) Heating effects caused by power dissipation in the high-field region which causes the breakdown voltage to increase: this effect can be represented by an effective electrical resistance R_{th}.

In a well-designed avalanche diode the parasitic resistance is much smaller than either the space-charge resistance or the thermally generated resistance. Thus R_p does not play a significant role in determining the shape of the d.c. current–voltage characteristic. The microwave performance of the oscillator is, however, significantly affected by the parasitic resistance, and this will be discussed in detail in Chapter 5.

We shall consider first the space-charge effects [7] by assuming that the temperature of the junction is constant. With a current I flowing in the device, the voltage V across the diode terminals is given (assuming $V > V_{BO}$) by

$$V = V_{BO} + IR_{sc}, \tag{3.22}$$

where V_{BO} is the room-temperature value of breakdown voltage as calculated above. The space charge of the carriers distorts the electric field profile in the space-charge layer, giving rise to a positive d.c. incremental resistance. To calculate the effects of this space charge let us consider an idealized Read-type diode in which the carriers are generated at one edge of the space-charge layer of width w. We assume, further, that the field is large enough that the carriers are travelling throughout with the saturated drift velocity v_s. Then the carrier density in the space-charge region is given by

$$n = \frac{J}{qv_s}. \tag{3.23}$$

The electric field $\Delta E(x)$ produced by these charges is obtained from a linear integration of Poisson's equation, namely

$$\Delta E(x) = \frac{Jx}{q\epsilon v_s}, \tag{3.24}$$

and the voltage increase produced by the current flow is calculated by integrating across the layer. Thus the voltage increase ΔV above the breakdown voltage, produced by the current flow, is

$$\Delta V = \frac{Jw^2}{2\epsilon q v_s}, \tag{3.25}$$

and the space-charge resistance, defined in eqn (3.22), is given by

$$R_{sc} = \frac{w^2}{2A\epsilon q v_s}, \tag{3.26}$$

where A is the cross-sectional area of the diode.

Although eqn (3.26) was derived for a simple Read-type model it can be applied to more realistic structures such as the one-sided abrupt-junction diode where, as we have seen, the space-charge layer can be divided into separate avalanche and drift zones. In this cases the resistance of the drift zone is obtained from eqn (3.26) by substituting $(w_B - x_A)$ for w where w_B is the total space-charge layer width at breakdown and x_A is the width of the avalanche region.

For many practical devices, particularly those made from Si and GaAs, the effects of heating outweigh the space-charge effects, and the current–voltage characteristics in breakdown are determined largely by the temperature coefficient of the breakdown voltage and the thermal impedance of the diode. In fact, as we shall see later, this provides a convenient, non-destructive technique for measuring the diode thermal impedance and assessing its power-handling capacity. The decreased ionization rate at high temperature gives rise to a positive temperature coefficient of breakdown voltage, and in fact a reasonably good approximation is a linear dependence of voltage on temperature [8],

$$V_B(T) = V_{BO}[1 + \beta_0(T - T_o)], \tag{3.27}$$

where T_0 is room temperature and β_0 is the normalized temperature coefficient of the breakdown voltage

$$\beta_0 = \frac{1}{V_{BO}} \frac{dV_B(T)}{dT}. \tag{3.28}$$

As a result of this linear dependence the isothermal current–voltage characteristic shifts, without changing shape, to higher voltages as the temperature is increased. This is illustrated in Fig. 3.19. We have assumed that the space-charge resistance is not dependent on temperature. The actual I–V curve, under d.c. bias, reflects the rising temperature in the active region as shown by the solid line in Fig. 3.19. The I–V relationship is

$$I = \frac{1}{R_{sc}}\left[V - V_{BO}\{1 + \beta_0(T - T_0)\}\right], \tag{3.29}$$

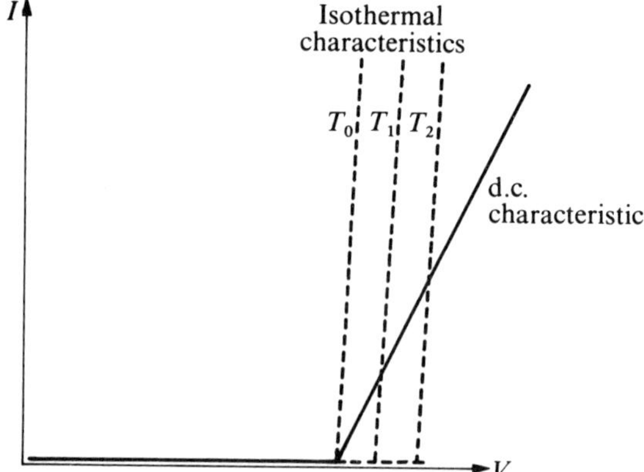

Fig. 3.19 Isothermal and d.c. voltage–current characteristic of reverse-biased junction.

and the voltage rise ΔV above the room temperature breakdown voltage is given by

$$\Delta V = V - V_{\mathrm{BO}} = I(R_{\mathrm{sc}} + R_{\mathrm{th}}), \tag{3.30}$$

where

$$R_{\mathrm{th}} = \frac{V_{\mathrm{BO}}}{I}\,\beta_0(T - T_0) \tag{3.31}$$

is the effective electrical resistance produced by the internal temperature rise at the junction. The thermal impedance θ of the diode is defined as the temperature rise produced at the junction by the thermal dissipation of one watt of power. That is

$$\theta = \frac{T - T_0}{VI}, \tag{3.32}$$

and from eqn (3.31) and (3.32)

$$R_{\mathrm{th}} = V_{\mathrm{BO}}\,V\beta_0\theta. \tag{3.33}$$

Measurements of R_{th} and β_0 enable the diode thermal impedance to be determined.

Typical values of R_{p}, R_{sc}, and R_{th} for a Si one-sided abrupt junction IMPATT at 10 GHz are 1 Ω, 50 Ω, and 200 Ω respectively, and in most practical devices the heating effect is dominant.

References

1. Shockley, W. The theory of p–n junctions in semiconductors and p–n junction transistors. *Bell Syst. tech. J.* **28**, 435 (1949).
2. Sze, S. M. and Gibbons, G. Avalanche breakdown voltages of abrupt and linearly graded p–n junctions in Ge, Si, GaAs, and GaP. *Appl. Phys. Lett.* **8**, 111 (1966).
3. Geraghty, S. R. Unpublished work.
4. Read, W. T. A proposed high-frequency, negative–resistance diode. *Bell Syst. tech. J.* **37**, 401 (1958).
5. Gibbons, G. and Sze, S. M. Avalanche breakdown in Read and pin diodes. *Solid-St. Electron.* **11**, 225 (1968).
6. Scharfetter, D. L. Power–Frequency characteristic of the TRAPATT diode mode of high-efficiency power generation in germanium and silicon avalanche diodes. *Bell Syst. tech. J.* **799** (May–June 1970).
7. Sze, S. M. and Shockley, W. Unit-cube expression for space charge resistance. *Bell Syst. tech. J.* **46**, 837 (1967).
8. McKay, K. G. Avalanche breakdown in silicon. *Phys. Rev.* **94**, 877 (1954).

4 IMPATT mode—principles of operation

4.1 Introduction

A device posesses negative resistance when the a.c. current lags the voltage by a phase angle between 90° and 270°. The negative resistance in an avalanche diode occurs as a result of a 180° phase difference developed between the a.c. current and voltage in a p–n junction reverse-biased into avalanche breakdown. The phase difference is produced by the time delay inherent in the build-up of the avalanche current, coupled with the phase delay developed as the carriers traverse the depletion layer. We shall examine below the origin of the phase delay in both of these cases by considering a simplified model for the device, in which the space-charge effects of the mobile carriers are ignored. By so doing, we are neglecting any distortion of the electric field profile within the structure by the space charge of the moving carriers, and assuming that the electric field within the device is in phase with the voltage across its terminals. In order to illustrate the origin of the negative resistance we shall consider the case where a sinusoidal voltage variation is impressed at the device terminals and examine the time dependence of the current in the diode. The starting conditions for oscillation will not be considered, since it is sufficient to demonstrate that the device possesses a negative resistance in order to account for the a.c. power generation. In a practical oscillator the diode is embedded in a circuit which is resonant at a frequency within the negative-resistance band of the device. The oscillation will then be started by random noise fluctuations, which will grow rather than decay in a negative-resistance medium, and oscillations will build up at the resonant frequency of the circuit.

4.2 Transit-time delay

Consider first the simple structure in which a semiconductor sample of length l, assumed to be depleted of electrons, is contacted at either end by metal electrodes. We assume that a high field exists within the sample, greater than the value required to produce saturation of the electron drift velocity. If at time $t = t_1$ an electron bunch is injected into the high-field region, the electrons will drift across the sample to the anode, where they will be collected at time

$t = t_1 + \tau$, where τ is the electron transit time ($\tau = l/v_s$). We are interested in the current induced in the external circuit. When the electrons are injected into the high-field region, charges are induced on both electrodes and a current will start to flow in the external circuit. As the electrons drift across, under the influence of the field, the induced charges on the electrodes vary, and so the current continues to flow until the electrons move into the anode, when the induced charges disappear. So the current in the circuit starts instantaneously when the charge is injected, but still flows up to τ seconds later. Thus, on average, the current flowing in the external circuit is delayed by $\tau/2$ relative to the current injected at the contact.

If a sinusoidal time-varying field is superimposed on the steady field in such a way that the amplitude is small enough for the electrons to remain at the saturation velocity, the electrons will again drift at a constant velocity, and hence a constant current will flow in the circuit even though the voltage is not constant. Fig. 4.1 illustrates three examples of the phase delay produced by the

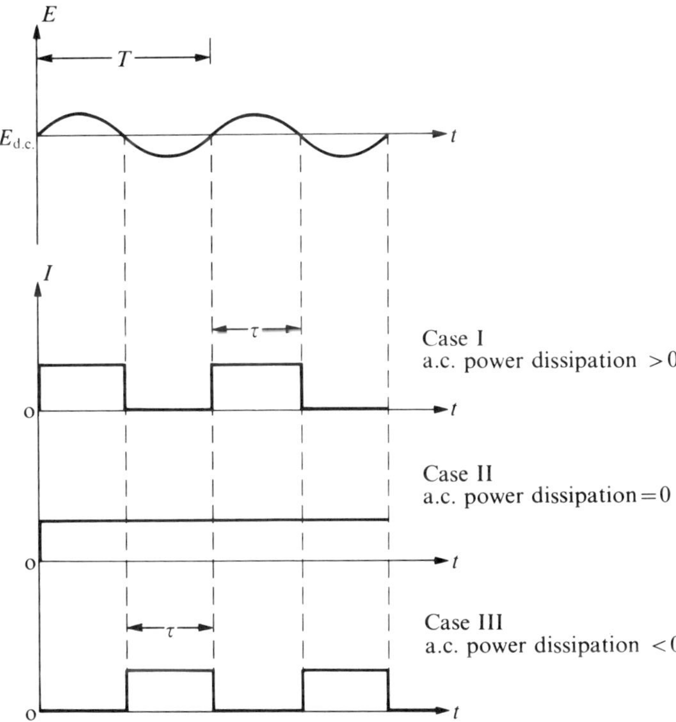

Fig. 4.1 Transit-time delay: three cases illustrating positive, zero, and negative a.c. power dissipation. In cases I and III the sample length is chosen such that the transit time τ is equal to one half of the period of the a.c. field swing.

transit-time effect, which lead to positive, zero, and negative a.c. power dissipation respectively. In case I, charge is injected into the drift space every cycle in phase with the voltage, and flows for a half cycle when the voltage is positive. The a.c. power dissipation is positive during the positive voltage half-cycle and zero during the negative half-cycle, so the net a.c. power dissipation per cycle is positive. In case II, the positive a.c. power dissipation, when the voltage is positive, is exactly balanced by the negative power dissipation when the a.c. voltage is negative, producing zero net a.c. power dissipation per cycle. In case III, by delaying the carrier injection one half-cycle so that current flows only when the a.c. voltage is negative, the power dissipation is zero during the positive half of the voltage cycle and negative during the negative half-cycle. This case corresponds to a.c. power generation. The length of the drift space in cases I and III is such that the carrier transit time is half the period of the a.c. cycle, so that on average the current flowing in the external circuit is delayed relative to the injected current by 90°. Case III is the one of interest here. The delayed injection, which is essential for a.c. power generation, is obtained by a 90° phase lag between current and voltage produced in the avalanche region of the diode. We shall show that this phase delay in the avalanche zone is an intrinsic property of the avalanche multiplication process. Of course, when power is generated in the a.c. circuit, as in case III, we are not contravening the first law of thermodynamics. It can be seen that the device will absorb power from the biasing (d.c.) circuit.

4.3 Avalanche delay

When a p–n junction is reverse-biased into avalanche breakdown, the current builds up from the thermally generated saturation current by the process of impact ionization, as described in Chapter 3. This current takes a finite time to build up after the application of the voltage, since the carriers must travel across the avalanche zone at the saturated drift velocity, producing ionizing collisions as they go. If a diode is d.c.-biased at the breakdown field and a sinusoidal time-varying voltage is superimposed on the d.c. level, then the current will build up during the half-cycle when the voltage is above the breakdown value and will decay when the voltage drops below the breakdown value, as illustrated in Fig. 4.2. Thus the current generated in the avalanche zone peaks one quarter cycle (90°) after the voltage has peaked, and since the carrier generation rate at any instant is proportional to the carrier density present, the current builds up exponentially as illustrated. This delayed current pulse is then injected into the drift zone. In this way the avalanche region acts as an injecting cathode producing a pulse of current, each cycle, which lags the voltage by 90°.

The time dependence of the avalanche multiplication process is readily derived from the continuity equations for holes and electrons. By considering a small volume within the avalanche zone, one may express the rates of change of electron and hole concentrations within this volume by the continuity equations

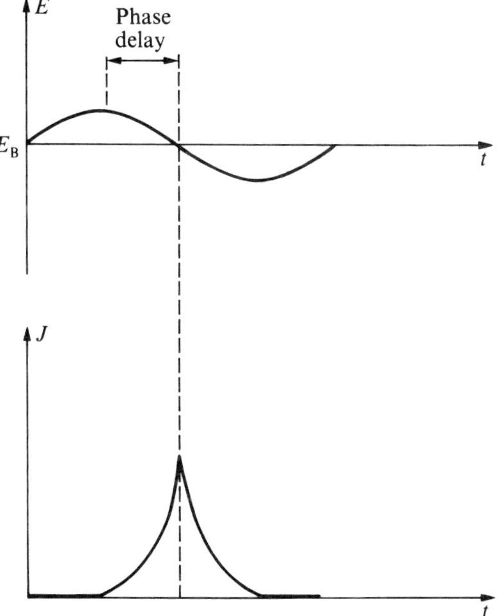

Fig. 4.2 Phase delay between current and field in avalanche zone

$$\frac{\partial n}{\partial t} = \frac{1}{q}\frac{\partial j_n}{\partial x} + \alpha v_s(n+p), \tag{4.1}$$

$$\frac{\partial p}{\partial t} = -\frac{1}{q}\frac{\partial j_p}{\partial x} + \alpha v_s(n+p), \tag{4.2}$$

where it has been assumed that holes and electrons have equal drift velocities and equal ionization rates. The first term on the right-hand side of (4.1) represents the net flux of electrons into the volume carried by the electron current j_n. The second term on the right-hand side of (4.1) represents the electron concentration generated within the volume by ionizing collisions, initiated by both electrons and holes within the volume. A similar interpretation holds for the hole continuity equation (4.2). The total particle current J in the avalanche zone is given by

$$J = j_n + j_p. \tag{4.3}$$

An expression for the time rate of change of the particle current in the avalanche zone is obtained by adding (4.1) and (4.2) and integrating over the avalanche zone:

$$\int_0^{x_A} \frac{\partial}{\partial t}(n+p)\,dx = \int_0^{x_A} \frac{1}{q}\frac{\partial}{\partial x}(j_n - j_p)\,dx + 2\int_0^{x_A} \alpha v_s(n+p)\,dx.$$

Using $j_n = qv_s n$ and $j_p = qv_s p$ and performing the integration, we obtain

$$\tau \frac{\partial J}{\partial t} = \Big[j_n - j_p \Big]_0^{x_A} + 2 \int_0^{x_A} \alpha \, dx,$$

where $\tau = x_A/v_s$ is the carrier transit time across the zone. At $x = 0, j_p$ is equal to the thermally generated hole saturation current j_{ps} and so $j_n - j_p = J - 2j_{ps}$. Similarly at $x = x_A, j_n - j_p = 2j_{ns} - J$; then the above equation reduces to

$$\frac{\tau}{2} \frac{\partial J}{\partial t} = J \left(\int_0^{x_A} \alpha \, dx - 1 \right) + J_s, \tag{4.4}$$

where $J_s = j_{ns} + j_{ps}$. In deriving (4.4) we have assumed that J is independent of x. Although this is certainly true in a static analysis it is not quite exact in the dynamic case, but we have used this approximation here, following Read [1], in order to simplify the integration. For the case of equal ionization rates the condition for static breakdown is given by (3.19).

$$\int_0^{x_A} \alpha \, dx = 1,$$

which defines the static breakdown field E_B. When a field $E > E_B$ is applied, $\int_0^{x_A} \alpha \, dx > 1$, the right-hand side of eqn (4.4) is positive, and the avalanche current J builds up. When the field drops below the static breakdown value, $\int_0^{x_A} \alpha \, dx < 1$, the right-hand side of eqn (4.4) becomes negative, and the avalanche current decays. This effect is illustrated in Fig. 4.2.

4.4 Oscillator efficiency

The d.c.-to-a.c. conversion efficiency of the diode is readily calculated, assuming a sinusoidal voltage and a current waveform as shown in Fig. 4.1(c), by Fourier analysing the square-wave current and considering the fundamental frequency only. For a fully modulated square wave, shown in Fig. 4.1(c), the a.c. current amplitude is equal to the d.c. value and the fundamental Fourier component is $4/\pi \, I_{d.c.} \sin \omega t$. Then the conversion efficiency η is given by

$$\eta = \int_0^T \frac{4 I_{d.c.}}{\pi} \sin \omega t \, V_{a.c.} \sin \omega t \, dt / I_{d.c.} V_{d.c.} \tag{4.5}$$

$$= \frac{2}{\pi} \frac{V_{a.c.}}{V_{d.c.}}, \tag{4.6}$$

and depends on the maximum allowable a.c. voltage swing $V_{a.c.}$ on the diode.

This is determined by large-signal limitations in the drift space. If the field swings above the breakdown value, impact ionization will take place in the drift zone. On the other hand, if the field swings below the value required to maintain the saturated drift velocity, the carrier bunch will collapse. In either case the phase relationship between current and voltage is destroyed. Read [1] has suggested that a realistic limit on the field swing in the drift space is an a.c. amplitude one-half of the d.c. field,

$$E_{\text{a.c.}} = \frac{E_{\text{d.c.}}}{2}.$$

If we assume that the avalanche region of the diode is infinitely thin, the applied voltage is dropped entirely across the drift space and the maximum a.c. voltage amplitude is then.

$$V_{\text{a.c.}} = \frac{V_{\text{d.c.}}}{2}.$$

Therefore, with this approximation, the maximum efficiency from (4.6) is

$$\eta = \frac{1}{\pi} \approx 30 \text{ per cent.}$$

This is an optimistic result and in practice the maximum efficiency obtained has been about 15 to 20 per cent. The discrepancy arises primarily from assuming that the avalanche zone is infinitely thin. In fact, power is dissipated in the avalanche region of all practical devices, and to a large extent the power used to create the carriers is wasted since the a.c. power generation takes place predominantly in the drift space. We will return to this point in Chapter 5, when the effect of the extended avalanche zone on the oscillator efficiency is considered in some detail. There we will show that the width of the avalanche zone depends on the ratio of the ionization rates (α/β).

4.5 Small-signal impedance

The impedance of the Read diode under the assumption of small a.c. amplitude was first calculated by Read [1] and subsequently extended by Gilden and Hines [2], whose approach we will follow here. The impedance of the avalanche and drift zones are calculated separately, assuming that the current and field are each composed of a d.c. part and a small a.c. variation; using the conventional small-signal approximation, products of a.c. terms are neglected wherever they appear. Thus

$$J = J_0 + j_a \, e^{i\omega t},$$

$$E = E_0 + \mathcal{E}_a \, e^{i\omega t}, \tag{4.7}$$

where J_0 and E_0 are the d.c. values of current and field respectively and j_a and $\mathcal{E}_a$ are the amplitudes of the a.c. parts. We shall consider a diode of unit area.

Avalanche zone impedance

The impedance of the avalanche region is calculated from the expression relating the rate of current build-up to the ionization rate (4.4) by integrating over the avalanche region subject to the small-signal assumptions. We shall neglect the saturation current for simplicity and rewrite (4.4) to obtain

$$\frac{dJ}{dt} = \frac{2J}{\tau}\left(\int_0^{x_A} \alpha \, dx - 1\right). \tag{4.8}$$

Assuming a spatially uniform field in the avalanche zone of width x_A, this reduces to

$$\frac{dJ}{dt} = \frac{2J}{\tau}(\alpha x_A - 1). \tag{4.9}$$

The ionization rate can also be divided into a d.c. part and a small a.c. part proportional to the a.c. field. Therefore

$$\alpha = \alpha_0 + \alpha_a e^{i\omega t} = \alpha_0 + \frac{d\alpha}{dE} \mathcal{E}_a e^{i\omega t}, \tag{4.10}$$

and so

$$\alpha x_A = 1 + \frac{d\alpha}{dE} x_A \mathcal{E}_a e^{i\omega t}.$$

A relationship between the a.c. current and field is obtained by substituting (4.7) and (4.10) into (4.9). Selecting only the first-order a.c. terms we obtain

$$j_a = \frac{2\left(\frac{d\alpha}{dE}\right) x_A J_0 \mathcal{E}_a}{i\omega\tau} \tag{4.11}$$

for the a.c. particle current in the avalanche zone. The impedance per unit area of the avalanche region $(x_A \mathcal{E}_a)/j_a$ is given by

$$Z_A = (i\omega\tau)/\left[2\left(\frac{d\alpha}{dE}\right)J_0\right]$$

and is purely reactive; Z_A is in fact the impedance of an inductance of magnitude L_A given by

$$L_A = \tau/\left[2\left(\frac{d\alpha}{dE}\right)J_0\right].$$

This inductance is in parallel with the depletion-layer capacitance of the avalanche zone

$$C_A = \frac{\epsilon A}{x_A},$$

so the total impedance is the parallel combination of the capacitance C_A and the inductance L_A. The particle current j_a flows through the inductance, and a displacement current j_d flows through the capacitance given by

$$j_d = i\omega\epsilon\,\mathcal{E}_a.$$

The displacement current, of course, is exactly 90° out of phase with the voltage on the capacitance and neither dissipates nor generates power. The resonant frequency of this parallel LC circuit is called the avalanche resonance frequency ω_a, and is given by

$$\omega_a = \left[\frac{2\left(\dfrac{d\alpha}{dE}\right)x_A J_0}{\epsilon\tau}\right]^{\frac{1}{2}}. \tag{4.13}$$

The total circuit current, both conduction current and displacement current, flowing into the device is

$$j = j_a + j_d.$$

Since the voltage developed across the capacitance C_A is the same as that across the inductance L_A, we find that

$$j_d = -\frac{\omega^2}{\omega_a^2}j_a, \tag{4.14}$$

and therefore the particle current in the avalanche zone expressed as a fraction of the total current is given by

$$\frac{j_a}{j} = \frac{1}{1 - \dfrac{\omega^2}{\omega_a^2}}. \tag{4.15}$$

The total impedance of the avalanche zone, from the parallel LC circuit, is given by

$$Z_A = \frac{i\omega\tau/2\left(\dfrac{d\alpha}{dE}\right)J_0}{\left(1 - \dfrac{\omega^2}{\omega_a^2}\right)}. \tag{4.16}$$

The drift-zone impedance

For convenience we shall drop the exponential term and consider amplitudes only. The total a.c. current j in the drift zone is the sum of the displacement current and the particle or conduction current injected from the avalanche zone

$$j = j_c(x,t) + j_d(x,t). \tag{4.17}$$

The conduction current j_c in the drift zone at time t and position x is equal to the injected particle current j_a at $x = 0$ at time $t = -x/v$, and is given by

$$j_c(x,t) = j_a(t - x/v).$$

Thus under the assumption of a constant drift velocity the a.c. conduction current in the drift zone propagates as an unattenuated wave and can therefore be represented by

$$j_c(x) = j_a\, e^{-i\omega x/v}. \tag{4.18}$$

Now the impedance of the drift space is given by

$$Z_D = \tilde{v}/j$$

where $\tilde{v}$ is the a.c. voltage across the drift zone and j is the total a.c. current. The a.c. drift field is given by

$$\mathcal{E}(x) = j_d/i\omega\epsilon$$

which becomes, using (4.17) and (4.18),

$$\mathcal{E}(x) = \frac{1}{i\omega\epsilon}\left(j - j_a\, e^{-i\omega x/v}\right). \tag{4.19}$$

The field is expressed in terms of the total circuit current j, using (4.15),

$$\mathcal{E}(x) = \frac{j\left[1 - \left(\dfrac{1}{1 - \dfrac{\omega^2}{\omega_a^2}}\right) e^{-i\omega x/v}\right]}{i\omega\epsilon}. \tag{4.20}$$

The voltage across the drift zone is obtained by integration of (4.20):

$$\int_0^{x_D} \mathcal{E}(x)\,dx \equiv \tilde{v} = \frac{x_D j}{i\omega\epsilon}\left[1 - \frac{1}{1 - \dfrac{\omega^2}{\omega_a^2}}\left(\frac{1 - e^{-i\omega x_D/v}}{i\omega\tau_D}\right)\right], \tag{4.21}$$

where $\tau_D = x_D/v$ is the carrier transit time across the drift zone. The impedance of the drift zone is obtained from (4.21) and is given by

$$Z_D = \frac{1}{i\omega C_D}\left[1 - \frac{1}{1 - \dfrac{\omega^2}{\omega_a^2}}\frac{\sin\theta_0}{\theta_0}\right] + \frac{1}{\omega C_D}\left[\frac{1}{1 - \dfrac{\omega^2}{\omega_a^2}}\frac{1 - \cos\theta_0}{\theta_0}\right], \quad (4.22)$$

where $\theta_0 = wx_D/v$ is called the transit angle and C_D is the depletion-layer capacitance.

The total impedance of the diode can now be obtained by combining the avalanche-zone and drift-zone impedance (4.16) and (4.22). We obtain

$$Z = \frac{i\omega\tau/2\left(\dfrac{d\alpha}{dE}\right)J_0}{(1 - \omega^2/\omega_a^2)} + \frac{1}{i\omega C_D}\left[1 - \frac{1}{1 - \dfrac{\omega^2}{\omega_a^2}}\frac{\sin\theta_0}{\theta_0}\right] + \frac{1}{\omega C_D}\left[\frac{1}{1 - \dfrac{\omega^2}{\omega_a^2}}\frac{1 - \cos\theta_0}{\theta_0}\right]. \quad (4.23)$$

The real part of the diode impedance

$$\mathrm{Re}(Z) = \frac{1}{\omega C_D}\left[\frac{1}{1 - \dfrac{\omega^2}{\omega_a^2}}\frac{1 - \cos\theta_0}{\theta_0}\right]$$

is positive for frequencies below the avalanche resonance frequency, $\omega < \omega_a$, and negative at frequencies above ω_a. From (4.23) we see that the diode reactance also changes sign at the avalanche resonance. For $\omega < \omega_a$ the diode is inductive and for $\omega > \omega_a$ the diode is capacitive. Thus the device possesses negative resistance at those frequencies where it is capacitive. Our analysis so far has included two basic simplifications. We have dealt exclusively with the idealized Read structure in which the avalanche zone is taken to be infinitely thin, and we have not considered the effects of carrier space charge. Before considering the extended avalanche model we shall examine these space-charge effects in the drift zone of the Read diode. Several important limitations on the performance of the oscillator are based on these effects.

4.6 Space-charge effects

The space charge of the electron bunch in the drift zone modifies the electric field profile as shown in Fig. 4.3. Behind the electron bunch the field is depressed and in front of the pulse the field is increased in such a way that the area under the field curve is unchanged. The change in field ΔE between the leading and trailing edges of the pulse is related to the total charge in the pulse Q by Gauss' law

$$\Delta E = Q/\epsilon,$$

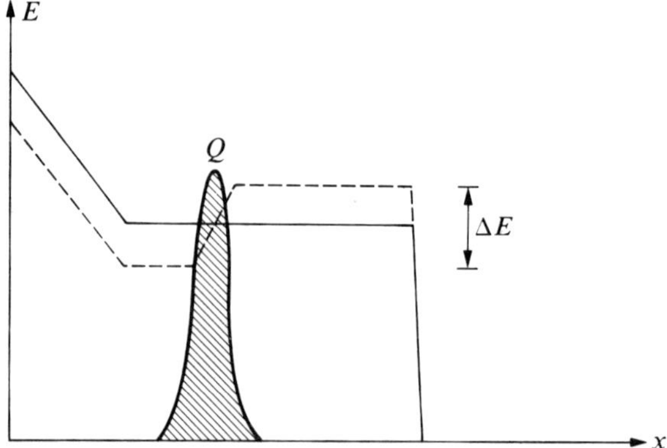

Fig. 4.3 Effect of carrier space charge on electric field profile.

where ϵ is the permittivity. As the electron bunch moves across the zone from left to right the amount of field depression behind the pulse decreases, but the field in front of the charge pulse steadily increases. This field distortion produces a deleterious effect in the avalanche zone as well as producing limitations on the power capability of the drift space.

The reduced field at the junction turns the avalanche off prematurely, as illustrated in Fig. 4.4 which shows the peak field in the avalanche zone, the voltage at the diode terminals, and the external current as functions of time. The field in the avalanche zone drops prematurely below the breakdown field at time t_1 and quenches the avalanche multiplication. However, the voltage waveform is not distorted by the redistribution of field within the structure and remains identical to the voltage waveform in the absence of space-charge effects. As a result, the current begins to flow when the a.c. voltage is positive, and a.c. power is dissipated for a short part of the cycle ΔT. In addition, since the electron transit time is one-half of the period of the a.c. cycle, the current turns off before the a.c. voltage returns to zero. Clearly the efficiency of the device is degraded. As the current level is increased the space-charge effects get more pronounced and the efficiency degradation gets worse.

In the drift space two effects are important. As the electron bunch approaches the anode the field in front of the pulse increases sharply due to the carrier space charge. This occurs during the part of the cycle when the a.c. field is increasing and approaching the d.c. level. To avoid impact ionization at this point in the structure, the field at the leading edge of the pulse must always be less than the breakdown value. Note that if ionization is allowed to take place at the anode, a pulse of holes will be produced which will then drift towards the cathode during the subsequent half-cycle and induce a current in the external circuit when the a.c. voltage is positive, causing a.c. power dissipation. An upper limit for the total charge Q in the electron pulse is obtained by setting the field

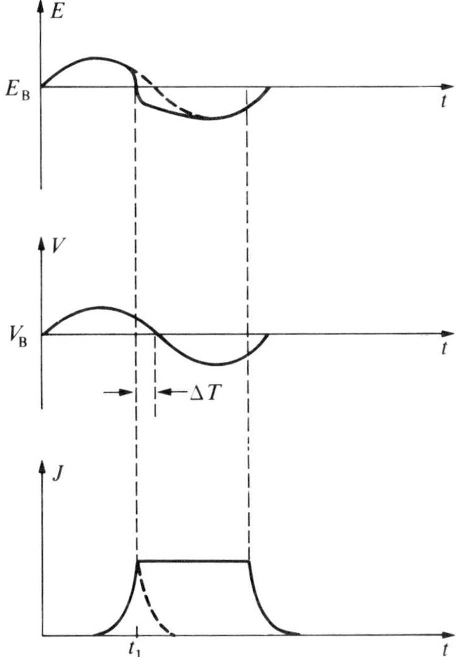

Fig. 4.4 Field, voltage, and current, including the effects of space charge.

rise at the anode ΔE, equal to the breakdown field E_B. Clearly this somewhat overestimates the maximum allowable charge, but we are assuming that the d.c. field in the drift zone is small compared with the breakdown field. Then by Gauss's law

$$Q_{max} = \epsilon E_B. \tag{4.24}$$

One electron pulse is produced each cycle and therefore the d.c. current density J_0 is given by

$$J_0 = Qf, \tag{4.25}$$

where f is the frequency of oscillation. Thus the maximum current density to prevent ionization in the drift space is, from (4.24) and (4.25),

$$J_0\,(\text{max}) = \epsilon E_B f. \tag{4.26}$$

For current densities greater than this value the avalanche diode will not oscillate in the IMPATT mode. In practical devices, at low frequencies (below about 50 GHz), the current density is limited by heating effects to values well below this critical value. However, at millimetre-wave frequencies the operating current density may approach or even be limited by the critical value. We shall return to this point in the next chapter when we derive a power–frequency relationship for the IMPATT oscillator.

The field depression behind the electron pulse in the drift zone may also degrade the performance of the device. The field must everywhere remain above the value required to produce velocity saturation for optimum operation in the IMPATT mode. The field depression is greatest as the electron pulse emerges from the avalanche region, but the minimum in the a.c. field swing occurs when the electron pulse is halfway across the drift space. Thus the position within the drift zone at which the field is a minimum depends on the relative magnitudes of the a.c. swing and the field modulation produced by the carrier space charge. If the field is allowed to drop below the saturation value, the electrons at the trailing edge of the pulse will slow down. The electron bunch will spread out and eventually collapse, destroying the phase relationship between the voltage and current in the external circuit.

4.7 Generalized small-signal analysis

The idealized model presented in §4.5, in which the avalanche region is assumed to be infinitely thin, is not a realistic representation for most devices. In most practical diodes the avalanche region occupies a significant fraction of the total active region, and a large part of the input power is used up in the avalanche zone to produce the holes and electrons. Misawa [3] and Gummel and Scharfetter [5] have analysed the extended-avalanche model using numerical techniques.

An approximate small-signal analysis has been presented by Misawa [6] using a multiple uniform-layer approximation, in which the diode active region is divided into regions of uniform field. In this model the one-sided abrupt junction (p^+–n or n^+–p) is approximated by a uniform-field avalanche zone of finite width in series with a uniform (but lower) field drift zone. The double-drift diode (p^+–p–n–n^+) is approximated by a uniform-field avalanche zone bounded on each side by a uniform (but lower) field drift zone. The small-signal impedance of any type of device can be calculated analytically, using this approximation, by solving the continuity equations and Poisson's equation in both the avalanche and drift regions separately. The solutions in each region are then matched by assuming continuity of electric field and current flow at the boundary. In this way the small-signal impedance may be obtained for various diode structures using the fraction of the depletion layer occupied by the avalanche zone as the principal diode design parameter. Before discussing the results of this analysis we must first examine more carefully the current–voltage relationship in a uniformly avalanching region.

Uniformly avalanching region

In the previous discussions the phase delay between current and voltage in the avalanche region was shown to be 90° in the case where the spatial extent of the avalanche zone was neglected. However, when an extended avalanche region is considered, a phase delay greater than 90° can in fact be developed between the current and voltage. That is to say, the avalanche region itself possesses a negative resistance. This negative resistance is distinctly different from that produced by the transit of carriers in the drift space of the IMPATT diode. It is smaller in magnitude, less sensitive to frequency, and in fact exists over a very wide frequency range.

Misawa [3] has studied in detail the dynamics of avalanche multiplication in a uniformly avalanching region (p–i–n diode). He has shown that, in such a medium, any perturbation will oscillate in time and propagate with a growing amplitude. The origin of this negative resistance, which is produced by the combined effects of avalanche delay, carrier drift, and space charge of the generated carriers, can be understood with reference to Fig. 4.5. Consider a fluctuation in the hole density as illustrated by the solid line in Fig. 4.5(a). Under the influence of the applied electric field the hole distribution will drift to the right at the saturated drift velocity. The space-charge fluctuation produces a fluctuation in the electric field shown by the dotted line. By Gauss's law the gradient of the electric field is proportional to the excess hole density. The generation rate of carriers by impact ionization is large where the field is high and where the carrier density is large, so that the maximum generation occurs somewhere between the point at which the field is highest and the point at which the hole density is largest as shown by the vertical lines. From the previous discussion of the inductive delay in the avalanche process we know that the generated carrier density lags the generation rate by 90°. As a result, the generated charge appears at the points of maximum generation one-quarter of a cycle later as shown in Fig. 4.5(b). At this point in time, both the hole and field distributions have advanced one-quarter wavelength to the right. The points of maximum carrier generation have also advanced one-quarter wavelength to the right and once again we expect the new generated charge to appear one-quarter cycle later. This is illustrated in Fig. 4.5(c), after one-half period has elapsed. At the end of the second quarter cycle the holes generated during the first quarter cycle have drifted through one-quarter wavelength and have arrived at the point where the new generated charge now appears. We see therefore that the carrier bunch grows as it moves downstream. Later points in the cycle are shown in Fig. 4.5(d) and (e). Thus, contrary to what might at first be expected, the avalanche multiplication process in this instance produces bunching rather than randomizing of the carrier distribution.

The phase difference between the field and current can be determined from Fig. 4.6, which shows the hole density and field at point A in Fig. 4.5 as functions of time. The hole density produced by the initial fluctuation is shown as p and the hole density produced by avalanche ionization is shown as Δp. We see that the hole density of the initial fluctuation lags the field by 90°, but the hole density produced by the avalanche multiplication lags the field by more than 90°. Since the hole current is in phase with the hole density, the total current lags the field by a phase greater than 90° and a negative resistance is produced. Fig. 4.7 also illustrates the phase relationships by use of a vector diagram.

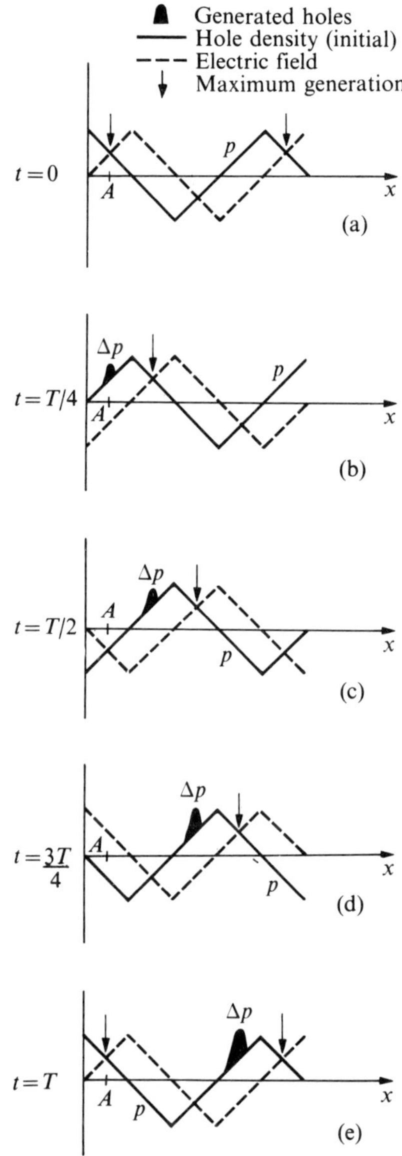

Fig. 4.5 Electric field and generated hole density in uniformly avalanching region resulting from an initial fluctuation in hole density (after Misawa [3]).

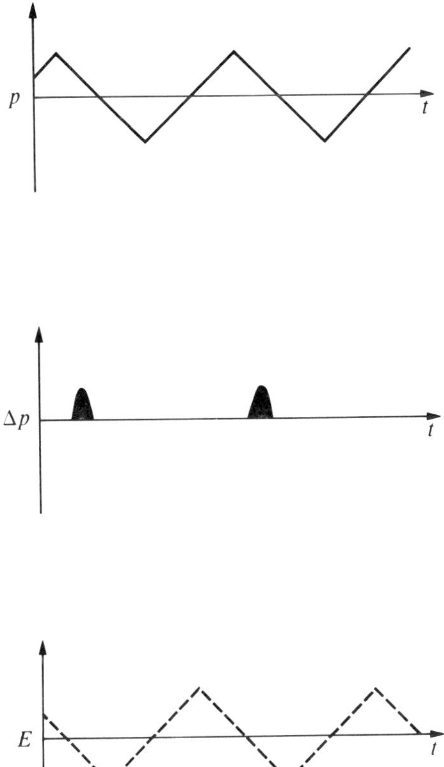

Fig. 4.6 Hole density (p, Δp) and electric field (E) as functions of time a fixed point (A) in uniformly avalanching region.

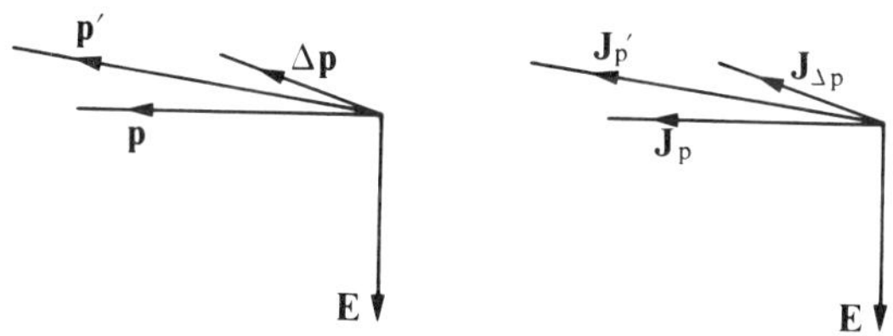

Fig. 4.7 Vector diagram illustrating phase relationship between current and field in uniformly avalanching region (after Misawa [3]).

Multiple uniform-layer approximation

The small-signal negative resistance, calculated by Misawa [6] using the multiple uniform-layer approximation, is shown in Fig. 4.8 for various structures as a function of frequency. The frequency is expressed as the transit angle defined in §4.5. The width of the avalanche zone, expressed as a fraction of the total space-charge layer width, is the parameter used to distinguish the various diode designs. Structure 1 has a very localized avalanche zone and approaches the Read model.

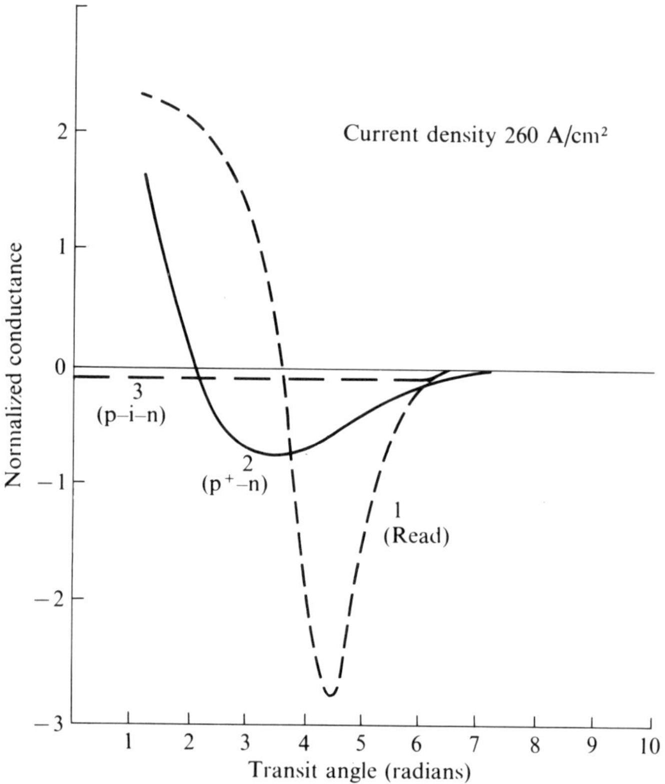

Fig. 4.8 Small-signal conductance of Read, p^+–n, and p–i–n diodes as a function of frequency (transit angle) (after Misawa [6]).

It is characterized by a large narrow band of negative conductance. Structure 3 is the p–i–n diode in which the negative resistance is that produced by the space-charge effects in the uniformly avalanching structure as discussed above. The negative conductance of this device is very small but extends over a very large range of frequencies. Between these two extreme cases lies the p^+–n (or n^+–p) diode in which the avalanche zone occupies about one-third of the depletion layer width. This device is easier to make than the Read diode and provides a reasonable compromise between negative resistance and bandwidth.

The small-signal theory describes quite well the characteristics of the diode when operated as a small-signal amplifier. However, because of the approximation that the a.c. signal level be small, this theory specifically excludes the large-signal oscillator case. The small-signal theory is, however, quite a useful guide in predicting, qualitatively, the properties of the oscillator.

Large oscillator output power is obtained when the a.c. voltage swing is large and when the negative resistance, at high signal levels, is large, the negative resistance being dependent on the a.c. voltage amplitude. The most useful small-signal parameter for predicting oscillator performance is the Q-factor of the diode. The Q-factor is defined [3] as the ratio of the energy stored in the device $\langle W_d \rangle$ to the power dissipated $\langle dW_d/dt \rangle$, times the angular frequency; that is,

$$Q_d = \frac{\omega \langle W_d \rangle}{-\langle \frac{dW_d}{dt} \rangle}. \tag{4.27}$$

The Q-factor of the diode is negative wherever it possesses negative resistance. In a similar way we may define the Q-factor of the circuit, which must be equal in magnitude to the diode Q for a stable oscillator. If the magnitude of the diode Q-factor is smaller than the circuit Q, oscillations will build up. Q_d decreases as the oscillation amplitude increases until the steady-state condition is reached. The growth rate of the oscillation g is given by

$$g = \left| \frac{1}{2Q_d} \right|, \tag{4.28}$$

and is large when the magnitude of the diode Q is small. Thus a small value of Q_d produces a rapid build-up of the oscillation. If the oscillation growth rate at small signal levels is large, it is to be expected that the oscillation will build up to a large amplitude and that the oscillator efficiency will be high. Thus we may interpret a small magnitude of Q_d as an indication of a high-efficiency oscillator. This interpretation should not be regarded as quantitatively exact, since the Q factor is defined under steady-state conditions and may not in every case accurately predict the transient behaviour of the device. However, it has been shown in practice to be a useful guide in predicting the microwave characteristics and operating conditions for different diode designs. Misawa [6] has calculated the diode Q-factor for the structures of Fig. 4.8. The results are shown in Fig. 4.9 for the Read and the p^+–n diodes at two values of current density. The smallest Q factor is obtained in the Read structure at low current densities. Again, however, the bandwidth is smallest, and as the avalanche zone is widened the magnitude of the diode Q increases and the band width increases.

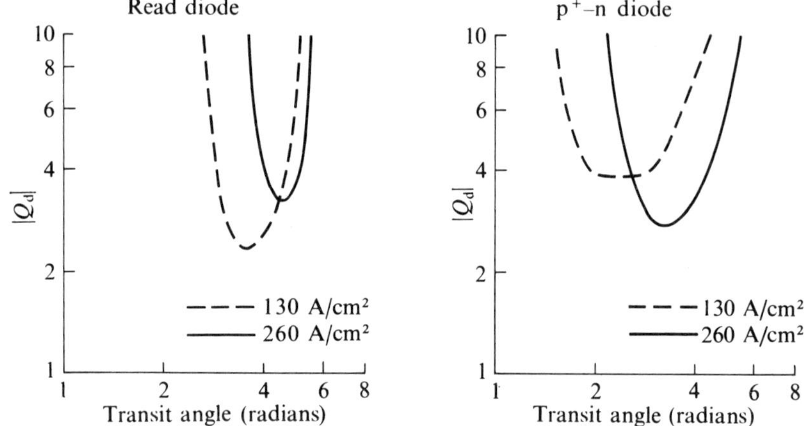

Fig. 4.9 Small signal Q of Read and p^{+}–n diodes as a function of frequency (transit angle) (after Misawa [6]).

From this we expect the highest efficiency oscillators to be those with the narrowest avalanche regions. One of the main disadvantages of the narrow avalanche structure lies in the fact that the Q is smallest at low-bias currents and degrades rapidly under high-drive conditions. It is not, therefore, the most suitable structure for high-power operation. The *pvn* structure is best for high-pulsed power operation, and the one-sided abrupt-junction diode (p^{+}–n or n^{+}–p) provides a good compromise between operating current density and favourable Q-factor for high-power CW oscillators.

4.8 Large-signal analysis

Theoretical calculations of the large-signal characteristics of the IMPATT oscillator cannot be performed analytically without making several rather drastic assumptions. Large-signal solutions to the equations describing carrier generation and space-charge balance can only be obtained, for realistic cases, by the use of numerical techniques. Scharfetter and Gummel [7] have presented a large-signal numerical solution for the Read diode oscillator, which describes the time evolution of the field, carrier densities, and current in the active region, and the voltage across the diode terminals. Using an implicit finite-difference method, self-consistent solutions are obtained to the basic time-dependent equations with realistic boundary conditions and circuit environment. A specific solution for a Read diode operating at 12·4 GHz, taken from Scharfetter and Gummel, is illustrated in Fig. 4.10. The electric field and carrier density throughout the structure are shown at four points in time during one complete cycle of oscillation. The 'snapshots' are one quarter-period apart. In the top left corner

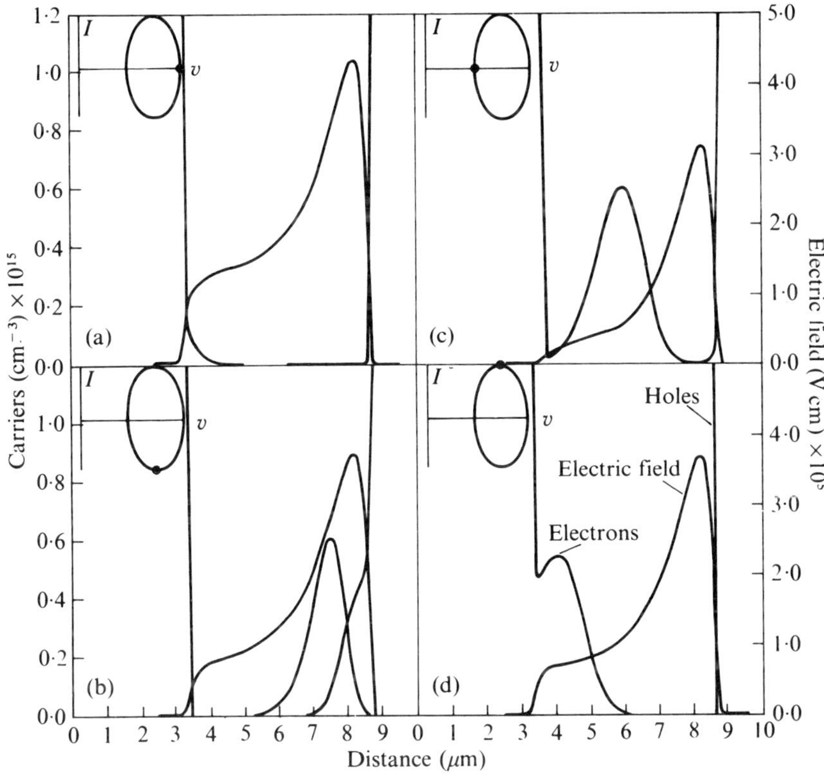

Fig. 4.10 Solutions of hole and electron concentrations, electric field, and terminal terminal current and voltage (values indicated by on phase plot) at various points in time for the diode operating at a frequency of 12·4 GHz, current density of 200 A/cm^2, and efficiency of 12 per cent. (After Scharfetter and Gummel [7]).

is shown the terminal current–voltage phase loop with the dot marking the appropriate point in time during the cycle. In Fig. 4.10(a) the a.c. voltage is at the peak of the swing, the avalanche has begun, but the generated charge density is immeasurably small. One quarter-cycle later (Fig. 4.10(b)), the voltage is passing through its mean value, and the hole–electron charge bunches have been generated and have begun to drift apart. After one half-cycle the hole bunch has almost disappeared on the right, the electron bunch has drifted to the centre of the structure, and the field and voltage have reached a minimum, as shown in Fig. 4.10(c). In the final picture (Fig. 4.10(d)) after three-quarters of the period has elapsed, the electron bunch has begun to disappear out the left-hand side of the region, and the field and voltage are both increasing and have already reached the mean value.

The detailed computations of Scharfetter and Gummel enabled the conductance and susceptance of the device to be evaluated under large-signal conditions and allowed the conversion efficiency of the oscillator to be calculated. However, because of the complexity of the problem, the computer time required to carry out a detailed device study is prohibitive, and as a result their study was limited to one particular Read diode structure. However, from this work, an approximate but simple design theory for the Read diode has been developed, based on a combination of small-signal theory and large-signal effects. From the small-signal theory the optimum growth rate of the oscillation occurs at a frequency approximately 20 per cent above the avalanche reäsonance frequency. This criterion is used to set up a relationship between the properties of the drift space and the avalance zone. The operating frequency of the oscillator, which is the drift frequency f_D given by

$$f_D = v_s/2x_D, \tag{4.28}$$

is chosen to be 20 per cent higher than the avalanche resonance frequency f_a, given by equation (4.13),

$$f_D \simeq 1{\cdot}2\, f_a. \tag{4.29}$$

Large-signal constraints are imposed on the current density by the field modulation produced as the charge pulse traverses the drift space. For a 50 per cent modulation of the field in the drift zone, the maximum charge in a single pulse is, by Gauss's law,

$$Q_{max} = \frac{\epsilon E_D}{2}, \tag{4.30}$$

and since one pulse transits each cycle, the d.c. current density is therefore given by

$$J = \frac{\epsilon E_D f_D}{2}. \tag{4.31}$$

Equations (4.29), (4.30), and (4.31) serve to define the structure and operating conditions for the Read-diode oscillator.

References

1. Read, W. T. A proposed high–frequency negative resistance diode. *Bell Syst. tech. J.* **37**, 401 (1958).
2. Gilden, M. and Hines, M. E. Electronic tuning effects in the Read microwave avalanche diode. *IEEE Trans. Electron Devices* **ED–13,** 169 (1966).
3. Misawa, T. Negative resistance in p–n junctions under avalanche breakdown conditions, Part I. *IEEE Trans. Electron Devices* **ED–13,** 137 (1966).

4. Misawa, T. Negative resistance in p–n junctions under avalanche breakdown conditions, Part II. *IEEE Trans. Electron Devices* **ED–13**, 143 (1966).
5. Gummel, H. K. and Scharfetter, D. L. Avalanche region of IMPATT diodes. *Bell Sys. tech. J.* **45**, 1797 (1966).
6. Misawa, T. Multiple uniform layer approximation in analysis of negative resistance p–n junction in breakdown. *IEEE Trans. Electron Devices* **ED–14**, 795 (1967).
7. Scharfetter, D. L. and Gummel, H. K. Large signal analysis of a silicon Read diode oscillator. *IEEE Trans. Electron Devices* **ED–16**, 64 (1969).

5 Performance limitations of IMPATT oscillators

In this chapter we shall be concerned with various phenomena that set limitations on the performance of the IMPATT oscillator. To cover in detail all aspects of the design limitations is beyond the scope of this book, and our discussion will be focused primarily on the principal effects which set limits on the efficiency, power, and frequency of the oscillator. Some of the limiting mechanisms to be discussed are inherent in the physics of the device itself. Others are set by factors external to the device, such as the circuit impedance limits and the ability to extract heat efficiently from the diode.

5.1 Inherent material factors affecting efficiency

We shall restrict this discussion to four of the most important factors, within the diode, which determine the efficiency of the oscillator. These are the effect of:

(a) *The ratio of ionization rates for holes and electrons.* It has been shown experimentally that materials such as Ge and GaAs, in which the ionization rates are equal or nearly equal, produce the highest efficiency IMPATTs. When the rates are not equal, as for example in Si, the avalanche zone is wider and the charge multiplication process is less efficient.

(b) *Saturation of ionization rates at high fields.* The saturation of α at high values of field leads to a degradation in the current waveform and efficiency of the oscillator at high frequencies.

(c) *Minority carrier storage.* In a p^+–n (or n^+–p) diode, back diffusion of the generated electrons (or holes) from the active layer into the neutral p^+ (or n^+) region will occur. These minority carriers will be stored in the neutral region while the remaining carriers are in transit, and will diffuse back into the active layer at a later time in the cycle, causing a premature avalanche which destroys the current–voltage phase relationship.

(d) *Parasitic resistance.* Microwave power, generated within the active layer, will be dissipated as heat in any positive parasitic resistance existing, for example, in the neutral regions of the semiconductor.

Effect of ionization rates

We have shown in §4.4 that the efficiency of the oscillator is given by $\eta = 2V_{\text{a.c.}}/\pi V_{\text{d.c.}}$ for the idealized Read model, in which avalanche zone is of negligible width. Let us consider now a structure with an extended avalanche zone, illustrated in Fig. 5.1, in which the field is assumed uniform in both the

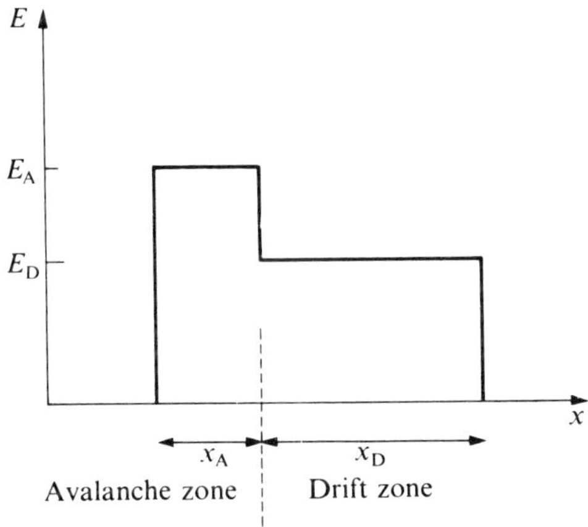

Fig. 5.1 Field profile of extended-avalanche model.

avalanche and drift regions. The a.c. voltage swing is limited by the maximum field swing allowed in the drift space, which we shall take, as before (§4.4), to be one-half of the d.c. field

$$E_{\text{a.c.}} = \frac{1}{2} E_{\text{D}}.$$

Hence

$$V_{\text{a.c.}} = \frac{1}{2}(x_{\text{A}} + x_{\text{D}}) E_{\text{D}},$$

and the efficiency for the extended avalanche model is

$$\eta \simeq \frac{1}{\pi}\left[\frac{1}{1 + V_{\text{A}}/V_{\text{D}}}\right], \tag{5.1}$$

provided the width of the avalanche zone is small compared with the width of the drift zone. $V_{\text{A}}/V_{\text{D}}$ is the ratio of the d.c. voltage drop in the avalanche zone to the d.c. voltage drop in the drift zone, and is a minimum when the ionization rates are equal. In Si, the electron ionization rate α is about 30 times larger than the hole ionization rate β, and as a result the avalanche zone in the p^+–n diode is about three times wider than for the equal ionization-rate case. This can be shown by considering the respective breakdown conditions (3.17) and (3.19) for both cases. If $\alpha = 30\beta$, the breakdown condition (3.17) becomes

$$\int_0^w \alpha \exp\left[\int_0^x 0{\cdot}97\, \alpha \, dx'\right] dx = 30,$$

and by making the substitution $y = \int_0^x \alpha \, dx$ it can be shown that

$$\int_0^w \alpha \, dx \simeq \frac{\ln 30}{0{\cdot}97} = 3{\cdot}4. \tag{5.2}$$

If we assume a uniform field in the avalanche zone, we obtain the width w by integrating (5.2),

$$w = 3{\cdot}4/\alpha_B, \tag{5.3}$$

where α_B is the value of the electron ionization rate at the breakdown field.

Now consider the case where $\alpha = \beta$ and $\alpha(E)$ has the same value as above. Then the breakdown condition (3.19)

$$\int_0^w \alpha \, dx = 1$$

can be integrated directly, for a uniform field, to give the width of the avalanche zone

$$w = 1/\alpha_B. \tag{5.4}$$

The wider avalanche region, when $\alpha \gg \beta$, can be understood with reference to Fig. 5.2. Consider first the case $\alpha = \beta$ shown in Fig. 5.2(a). An electron enters the avalanche zone from the left and travels through a mean free path before producing a hole–electron pair by impact ionization. The generated hole will then drift to the left, and, since $\beta = \alpha$, there is a high probability that this hole will suffer an ionizing collision and produce a hole–electron pair before it leaves the region. Thus the electron supply is continually replenished at the left-hand side of the zone, and the energy acquired from the field by both holes and electrons is efficiently used to create charge multiplication. However, when $\beta \ll \alpha$, as shown in Fig. 5.2(b), the holes created when the electrons suffer ionizing collisions are likely to leave the zone on the left without producing any new pairs. Thus, the energy acquired by the holes is not used, and charge multiplication is mostly produced by the electrons moving downstream. To produce the same amount of charge multiplication, the avalanche zone in this case must clearly be wider.

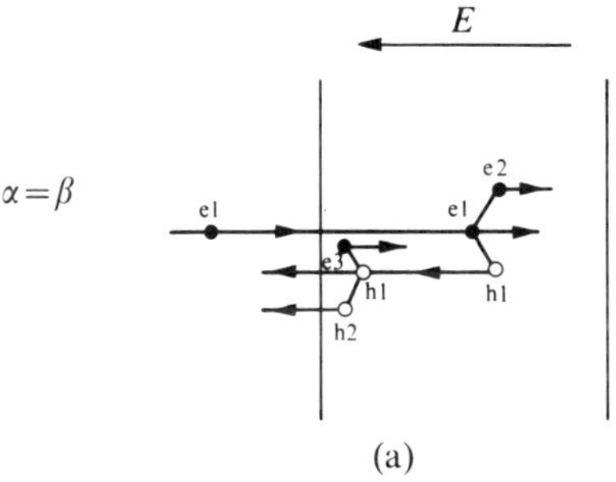

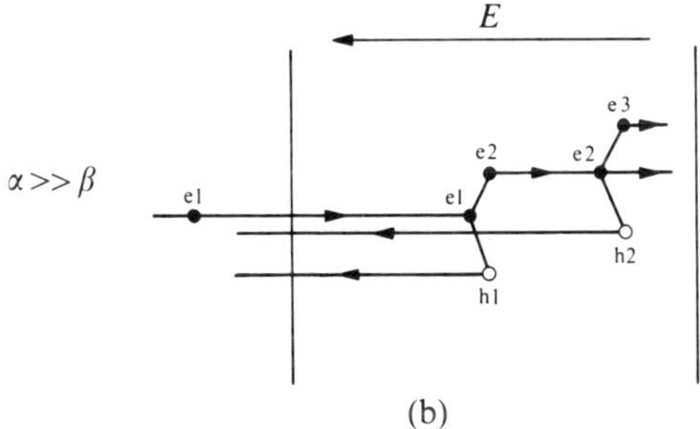

Fig. 5.2 Schematic of avalanche multiplication (a) when $\alpha = \beta$ and (b) when $\alpha \gg \beta$.

The ratio V_A/V_D is larger for the wider avalanche region, and from eqn (5.1) the oscillator conversion efficiency is smaller.

In a practical p^+–n (or n^+–p) IMPATT diode the field in the avalanche zone is not uniform, and if $\alpha \neq \beta$ the width of the avalanche zone depends on the polarity of the structure. When $\alpha \gg \beta$ the avalanche zone in the n^+–p diode is much smaller than in the p^+–n structure. This is the case for Si. This effect can be understood with reference to Fig. 5.3, which shows the electric field and current distribution in both polarity structures when $\alpha \gg \beta$ [9]. Since charge multiplication is large where both the field and the electron density are high, the avalanche is more localized in the n^+–p structure in which the electron density increases towards the metallurgical junction. For any semiconductor, the avalanche-zone width is smallest when the polarity is chosen so that the carrier with the highest ionization rate drifts towards the junction. When $\alpha = \beta$ the two polarities are identical. The improved efficiency expected from the narrow avalanche zone n^+–p Si IMPATT has been predicted in computer simulation experiments [10] and has recently been observed in experimental devices [11].

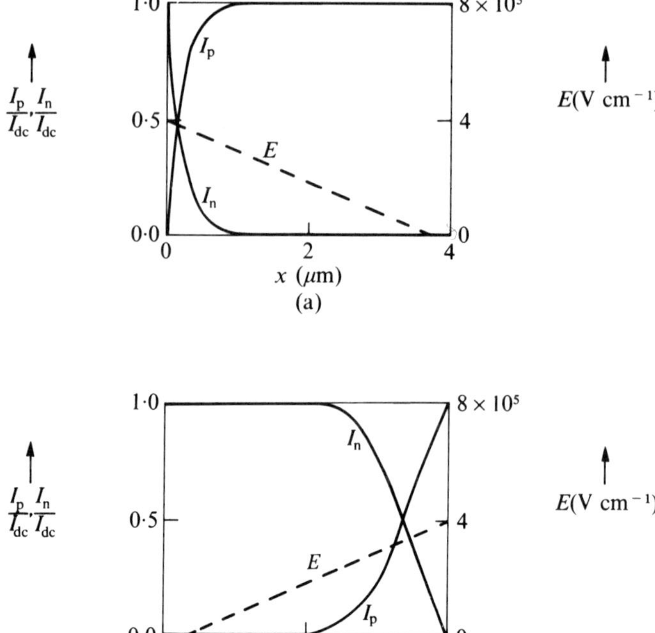

Fig. 5.3 Current and electric-field profiles for complementary Si one-sided abrupt junctions at breakdown
(a) n^+–p, $N_A = 7 \times 10^{15}$ cm^{-3}, $V_B = 76{\cdot}2$ volts.
(b) n–p^+, $N_D = 7 \times 10^{15}$ cm^{-13}, $V_B = 71{\cdot}4$ volts.
(after Schroeder and Haddad [9]).

Efficiency fall-off at high frequencies

The efficiency of the IMPATT diode decreases at high frequencies because of a saturation effect in the ionization rates at high values of field [1]. This fall-off in efficiency was not considered by Read [2] since he assumed a simplified relation between α and E of the form $\alpha = AE^m$, where A and m are constants. This approximation, though useful over limited field ranges, does not adequately describe the behaviour of the device at high frequencies since it predicts a constant fractional change in α, $\Delta\alpha/\alpha$, for a given fractional change in E, $\Delta E/E_B$, for all values of E_B. If the more accurate exponential relationship is used (eqn (2.19)), we find that $\Delta\alpha/\alpha$ is proportional to $1/E_B(\Delta E/E_B)$ in Si and to $1/E_B^2(\Delta E/E_B)$ in GaAs. This means that as the frequency is raised and the breakdown field increases, the fractional change in α decreases for a given fractional change in field about the breakdown value. As a result the response of the avalanche is slower and the current builds up and decays more gradually. The avalanche current waveform then becomes less pulse-like and assumes a more sinusoidal shape. Misawa [1] has calculated the reduction in efficiency, produced by this effect, in Si and GaAs p^+–n IMPATT diodes. The avalanche current and voltage waveforms calculated by Misawa for two Si diodes are shown in Fig. 5.4;

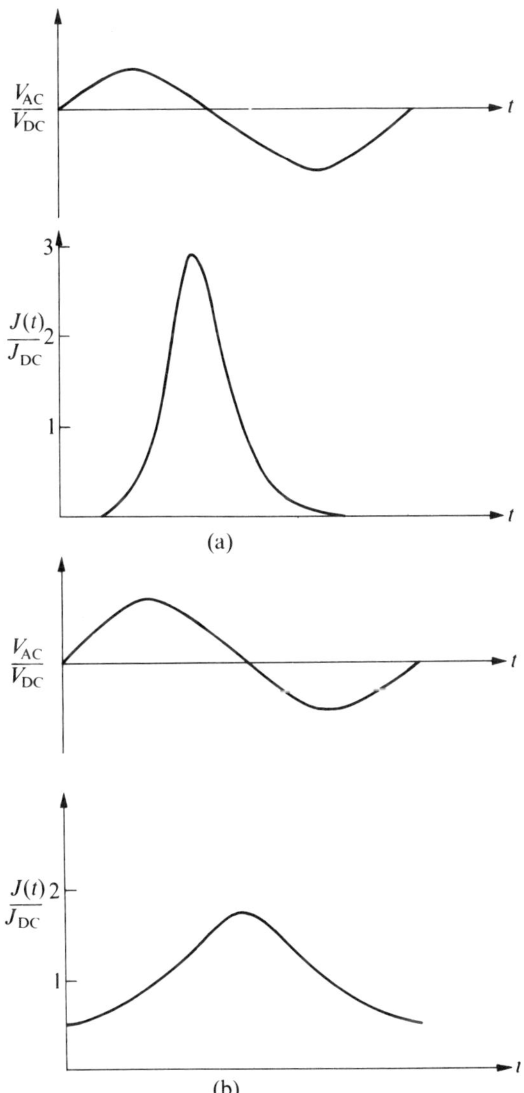

Fig. 5.4 Voltage and avalanche current waveforms in (a) 5·4 μm Si diode with $V_{a.c.}/V_{d.c.}$ = 0·44 and (b) 0·1 μm Si diode with $V_{a.c.}/V_{d.c.}$ = 0·38, showing the effect of saturation of the ionization rate on the shape of the current pulse (after Misawa [1]).

the first structure is a low-frequency diode with a width of 5.4 μm corresponding to an operating frequency of about 8 GHz; the second is a high-frequency diode with a width of 0·1 μm operating at a frequency above 300 GHz. The difference in the current waveforms is quite marked. The low-frequency diode turns on sharply and the current pulse is almost fully modulated. In the high-frequency unit the current builds up and decays more gradually and the swing is smaller. The fall-off in efficiency in narrow devices is shown in Fig. 5.5. These results are based on a maximum a.c. voltage amplitude of one-half of the d.c. voltage.

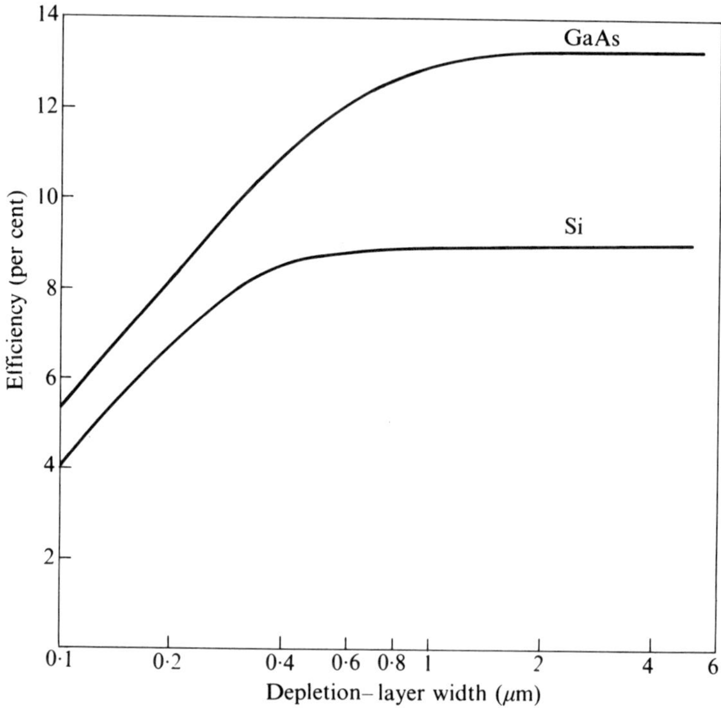

Fig. 5.5 Efficiency versus depletion-layer width in p^+–n IMPATT diodes showing the fall-off in efficiency in narrow devices (high frequencies) caused by the saturation of the ionization rate (after Misawa [1]).

Minority carrier storage

An important parameter affecting the efficiency of the oscillator is the magnitude of the reverse saturation current. Although the device is normally operated at bias currents many orders of magnitude greater than the saturation current, the diode nonetheless spends a considerable fraction of each cycle below breakdown, and the avalanche turn-on time is determined by the magnitude of the pre-breakdown current. As the field swings up through the breakdown value the carrier density builds up from the low residual level to a value determined by the bias current. The time required to build up to the final carrier density is critically dependent on the starting density, and a premature build-up of the avalanche current results when the saturation current is high. The phase delay associated with the avalanche process is reduced as shown in Fig. 5.6. The oscillator efficiency, for a square-wave current, is

$$\eta = \frac{2}{\pi} \frac{V_{\text{a.c.}}}{V_{\text{d.c.}}} \mid \cos \phi \mid, \tag{5.5}$$

which is similar to the expression given in (4.6) but includes a $\cos \phi$ term to take account of the phase angle between current and voltage. In the idealized Read model, $\phi = 180^\circ$ and $|\cos \phi| = 1$. When the saturation current is large, $\phi < 180^\circ$ and $|\cos \phi| < 1$.

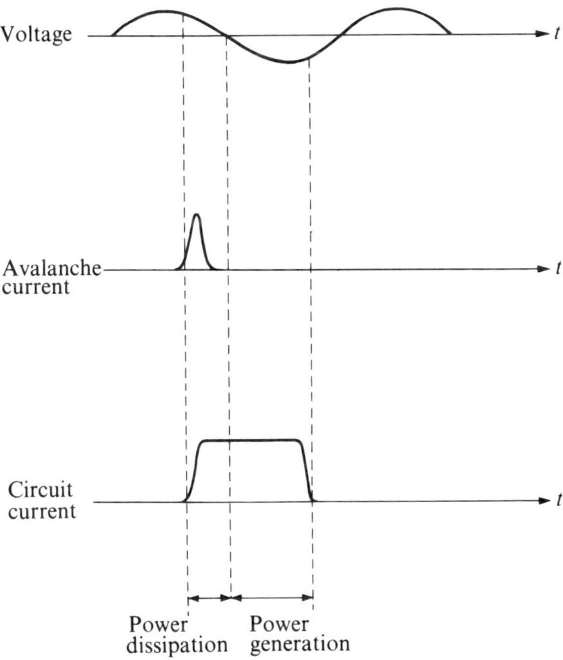

Fig. 5.6 Advance phase shift produced by excess saturation current.

Many factors may produce a large saturation current in the junction. One of particular importance is the minority-carrier storage effect produced by the back diffusion of carriers from the avalanche zone into the adjacent space-charge neutral region. These carriers are subsequently released into the active region and act as a very large saturation current to induce a premature build-uo of avalanche current. As the electron bunch builds up in the avalanche zone of a p^+–n junction, a very large electron concentration gradient is developed adjacent to the p^+ neutral region in which the electron concentration is very small. Some of the electrons diffuse back against the field into the p^+ region, where they remain as stored charge until the electron bunch in the avalanche zone moves away into the drift region. The electron concentration gradient at the p^+–n interface is then reversed and the stored electrons diffuse back into the active region inducing a premature build-up of the avalanche current. Misawa [3] has shown that this effect reduces the efficiency by about a factor of two. The charge storage effect, which is most pronounced in a Read structure, where the avalanche region is narrow, can be suppressed by using a Schottky barrier junction in place of the p^+–n junction since electrons cannot be stored in the metal.

Parasitic resistance

For efficient operation of an IMPATT oscillator it is necessary to transfer the power efficiently from the active region of the diode to the external load. In practice this is accomplished by matching the real part of the diode impedance to the load impedance, using a matching network with low loss at the frequency at which the total reactance of the circuit plus diode is zero (the resonant frequency). However, any losses within the diode remain unaffected by the external circuit and can, in some cases, severely degrade the overall conversion efficiency of the oscillator. If the device contains any positive resistance (for example in the contacts or neutral end regions), microwave power is dissipated there as heat. The simple equivalent circuit for the IMPATT diode shown in Fig. 5.7 is a negative conductance G, in parallel with the depletion-layer capacitance C. In series with this combination is a positive resistance element R_p, which represents the parasitic resistance of the device. When the oscillator

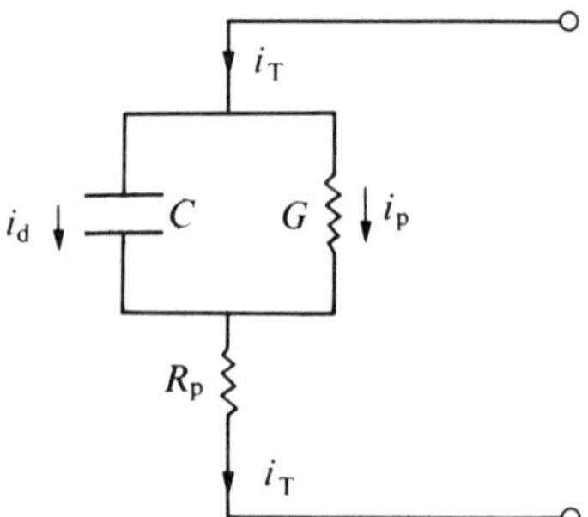

Fig. 5.7 Simplified equivalent circuit for IMPATT diode.

efficiency is high the d.c. current is fully modulated in the diode drift space as w have already shown in §4.4. Then the a.c. amplitude of the particle current is equal to the d.c. current.

$$i_p = I_{d.c.}$$

A large displacement current i_d flows in the depletion layer capacitance and is related, through the diode Q-factor, to the particle current

$$i_d = Q_d i_p.$$

The displacement current is 90° out of phase with the voltage across the capacitance, and neither dissipates nor generates power in the active region of the diode. Since $Q_d \gg 1$, the total a.c. current i_T flowing in the parasitic resistance is much larger than i_p. The generated microwave power is

$$P_{gen} = \frac{I_{d.c.}^2}{2G} = \frac{Q_d^2 I_{d.c.}^2 R}{2},$$

where R is the series equivalent negative resistance of the diode. The a.c. power dissipated in the parasitic resistance is

$$P_{dis} = \frac{Q_d^2 I_{d.c.}^2 R_p}{2},$$

so that the output power of the oscillator

$$P_{out} = P_{gen} - P_{dis},$$

is zero when the positive resistance is exactly equal in magnitude to the negative resistance of the device. This is expected since the diode then possesses no net negative resistance. Under large-signal conditions the negative resistance may be as low as a few ohms, and for efficient operation the diode must have a parasitic resistance less than one ohm.

Experimental confirmation of the critical role played by the parasitic resistance has been obtained for Ge diodes. In practical devices, for example a diffused n^+–p–p^+ structure, the main factors contributing to the positive parasitic resistance are the p^+ region and any part of the p region which remains undepleted at breakdown. In general, the n^+ diffused layer is very shallow and does not contribute a significant resistance. The p region is generally several orders of magnitude higher in resistivity than the p^+ contact region, and so the thickness of the p region must be tailored exactly to accommodate the active layer. Experimental results [4] on Ge IMPATT diodes, illustrating the effect of the undepleted expitaxial p-layer thickness on the oscillator efficiency, are shown in Fig. 5.8. For these devices, 1 μm of undepleted material produces a positive parasitic resistance of 1 ohm. A parasitic resistance of 2 ohms reduces the efficiency by about 50 per cent.

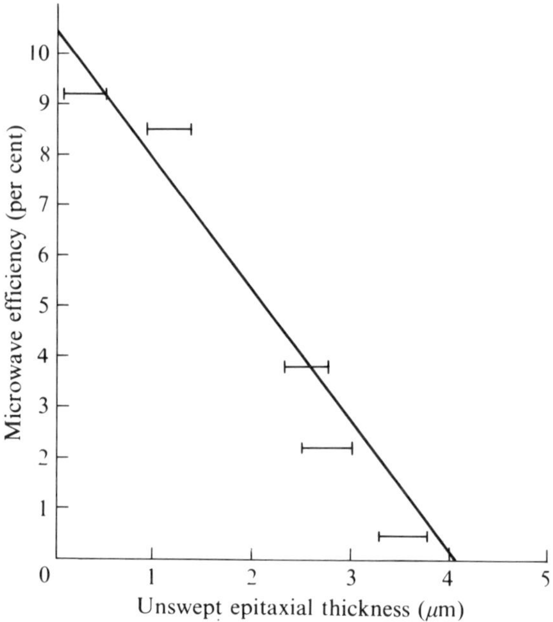

Fig. 5.8 Microwave efficiency versus undepleted epitaxial layer thickness for Ge IMPATT diodes (after Kovel and Gibbons [4]).

5.2 Transit-time limits

The ultimate performance of any device in which the frequency of operation is controlled by the carrier transit time in the active region is set by the well-known transit-time limitations. Avalanche-diode oscillators, Gunn oscillators, and transistors are all examples of transit-time limited devices in which the frequency of operation is inversely proportional to the width of the active region. As the frequency is increased, the width of the active region is reduced, and the voltages and power decrease accordingly. Thus, the transit-time limitation imposes a power–frequency trade-off given by an expression of the form

$$P_0 X_c f_0^2 = \text{constant},$$

where P_0 is the output power of the oscillator at frequency f_0 and X_c is the circuit impedance level. The constant depends on specific materials properties of the semiconductor. If the circuit impedance level is kept constant, the output power attainable from the oscillator decreases as $1/f_0^2$ and the maximum power at any frequency is obtained using the lowest realizable impedance level.

By considering the d.c. voltage and current limits we shall derive this power–frequency relationship for an IMPATT oscillator of the Read type with a localized avalanche zone and a uniform-field drift zone of width w. A similar argument can be developed for other transit-time devices [5, 6].

The maximum field in the active region of the device is limited by impact ionization to the breakdown value E_B. Then the maximum voltage that can be applied to the device is

$$V_m = E_B w. \tag{5.6}$$

The maximum current is set by the space-charge effects of the carriers moving through the drift space as discussed in Chapter 4. Thus, the maximum current is

$$I_m = \epsilon A E_B f_0,$$

or

$$I_m = wCE_B f_0, \tag{5.7}$$

where C is the capacitance of the diode which must be matched by the reactance of the circuit X_C:

$$X_c = \frac{1}{2\pi f_0 C}. \tag{5.8}$$

Combining eqns (5.6), (5.7), and (5.8), and using the relationship between frequency and diode width $f_0 = v_s/2w$, we obtain the maximum-power, frequency-squared product

$$P_m f_0^2 = \frac{E_B^2 v_s^2}{8\pi X_c}. \tag{5.9}$$

The diode impedance per unit area decreases as the frequency is raised, and the junction area must be decreased accordingly to bring the impedance up to a value which can conveniently be matched by a practical waveguide or coaxial circuit. At any frequency, the lower the circuit impedance level the more power the oscillator will deliver to the load. There are, of course, limits to the realizable impedance level in a practical microwave circuit. Read [2] has suggested a value of 10 ohms for the minimum impedance of an inductive waveguide cavity, and in practice this estimate appears to be a fairly realistic value for the lower limit.

It should be noted that the power discussed above is the d.c. input power to the device. From eqn (5.9) the two significant materials properties determining the power–impedance product are the breakdown field and the carrier saturated drift velocity. High-power capability results when both are large. In Chapter 2 it was shown that a large saturated drift velocity requires a high optical-phonon energy and small effective mass. In Chapter 3 it was shown that large breakdown fields are obtained in semiconductor materials possessing large energy-band gaps.

In the above analysis we considered a single-drift diode. An increased power–impedance product is obtained using the double-drift structure [7]. Since this device incorporates separate drift zones for both holes and electrons, the width of the active region is almost twice that of the single-drift diode at the same frequency. Then, for the same impedance, the area of the double-drift diode is larger by almost a factor of two and the maximum d.c. current is also twice as large. In addition the breakdown voltage is larger (see Chapter 3) and the net overall gain in the power–impedance product is about a factor of three. The full capability of the double-drift diode, under CW conditions, can only be realized in practice at high frequencies ($\geqslant 50$ GHz), since the maximum current density at low frequencies is determined by thermal limitations and not by the space-charge effects.

5.3 Thermal limitations

Since the efficiency of the IMPATT diode is relatively low, a large fraction of the d.c. power is dissipated as heat in the high-field region. The temperature at the junction rises above ambient, and in many practical cases the output power of the oscillator is limited by the rate at which heat can be extracted from the device. As the junction temperature increases, the reverse saturation current rises exponentially and eventually leads to a thermal runaway phenomenon resulting in the destruction of the device. Since, in contrast to the avalanche current, the reverse saturation current does not require a large voltage to sustain it, the voltage begins to decrease when the junction gets hot enough for the reverse current to constitute a significant fraction of the total current. A thermally induced d.c. negative resistance is produced, causing the current to concentrate in the hottest part of the diode. As well as leading to the eventual burn-out of the junction, the increased saturation current at elevated temperatures produces a degradation in the oscillator performance at power levels below the burn-out value. The increased reverse saturation current produces a faster build-up of the avalanche current and degrades the negative resistance of the device. Thus, in general, the oscillator efficiency will begin to decrease at power levels just below the burn-out power. The larger the band-gap of the semiconductor the smaller is the reverse saturation current and consequently the higher is the burn-out temperature of the junction. Thus, Ge IMPATTs ($E_g = 0{\cdot}7$ eV) are lower-power devices than either Si ($E_g = 1{\cdot}1$ eV) or GaAs ($E_g = 1{\cdot}43$ eV) devices.

The amount of d.c. power p_{max} that can be dissipated in a diode is determined by the burn-out temperature T_B and the thermal impedance of the device θ,

$$P_{max} = (T_B - T_0)/\theta, \tag{5.10}$$

where the thermal impedance is defined as the temperature rise produced at the junction by the dissipation of one watt of power. θ is determined by the size and shape of the diode and the thermal conductivity of the heat sink to which it is bonded. The heat is generated in the active region, and in a typical structure, illustrated in Fig. 5.9, it flows out through one side only. In the structure shown, the heat flows through the diffused region which is generally about 1 μm thick, through the metal contacts which are typically a few thousand ångströms thick, and into a semi-infinite copper heat sink. In most practical cases, especially at low frequencies where the junction diameter is large, the spreading resistance in the copper heat sink is the dominant contribution to the thermal impedance.

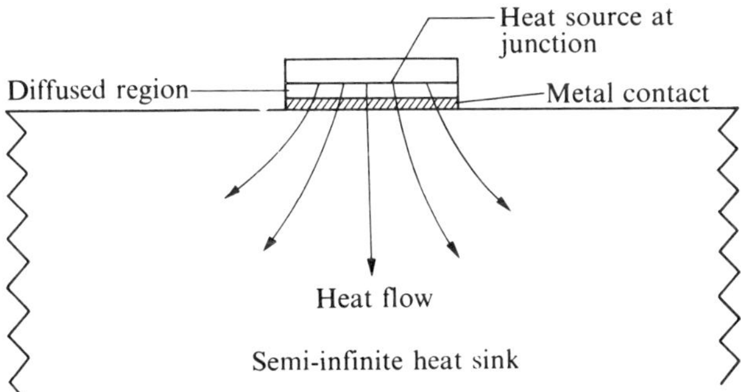

Fig. 5.9 Diode model for thermal impedance calculations.

The thermal impedance of any section is defined as the temperature difference across it produced by a heat flow of 1 watt. In the diffused region and in the contacts, the heat flow is linear and the thermal impedance is given by

$$\theta_c = \frac{L}{\pi r^2 K}, \tag{5.11}$$

where r is the diode radius, L is the width of the region, and K is the thermal conductivity. The spreading resistance in the heat sink is given by [12]

$$\theta_s = \frac{1}{\pi r K}, \tag{5.12}$$

if the heat flux is constant across the surface area of contact between the diode and the heat sink and by [12]

$$\theta_s = \frac{1}{4 r K}, \tag{5.13}$$

if the boundary between the contact and the heat sink is at constant temperature. In practice neither is exact. The centre portion of the diode is hotter than the edges, and since the breakdown voltage of the junction increases with temperature, the current distribution adjusts so that a higher current density flows through the edges of the diode. Thus neither the heat flux nor the temperature is uniform across the boundary. The actual spreading resistance, however, must lie somewhere between these two extreme values given by eqns (5.12) and (5.13). If we use the uniform-flux boundary condition the total thermal impedance of the diode is given by

$$\theta = \frac{L_s}{\pi r^2 K_S} + \frac{L_M}{\pi r^2 K_M} + \frac{1}{\pi r K_H}. \tag{5.14}$$

This thermal impedance is shown in Fig. 5.10, plotted against the diode area for the Si device illustrated in Fig. 5.9. The parameters used in the calculation are shown in Fig. 5.10. As the diode radius is decreased the relative magnitude of the contribution from the diffused region and the contacts increases. Notice that, in this calculation, we have assumed that the heat is produced in a plane at the junction, whereas in practice the heat is generated within the finite extent of the active region. This approximation simplifies the calculation and does not introduce a significant error except at very low frequencies where the active region is wide. Also shown in Fig. 5.10 is the total thermal impedance when the copper heat sink is replaced by type 11a diamond. Type IIa is the purest

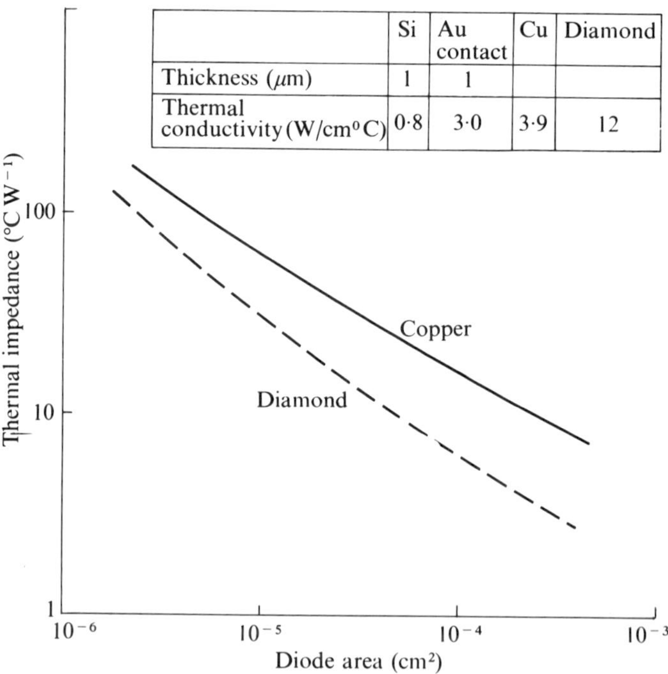

Fig. 5.10 Thermal impedance versus diode area for copper and diamond heat sinks.

form of diamond and has a thermal conductivity, at room temperature, about five times larger than that of copper. However, with increasing temperature, the thermal conductivity of diamond decreases as $1/T$ and at 500 K is only about three times that of copper. For the results shown in Fig. 5.10 we have chosen the values of thermal conductivity as those at 500 K, since this temperature is close to the operating conditions for practical oscillators. For Si and GaAs diodes thermal burn-out occurs when the junction temperature reaches approximately 300–350°C and a safe operating temperature rise is around 250°C. For an X band oscillator in Si and GaAs, the diode diameter is normally 100–150 μm and the maximum power that can be dissipated on a copper heat sink is 20 watts. Using a diamond heat sink, this power level can be raised to 55 watts.

5.4 Power–frequency limits for abrupt-junction IMPATTs

Based on the foregoing principles we are now able to establish a power–frequency relationship for practical abrupt-junction IMPATTs. We shall present a simplified analysis for Si and GaAs diodes employing the following assumptions:

(a) The diode thermal impedance θ is taken as the spreading resistance in a copper heat sink given by

$$\theta = \frac{1}{\pi r K}.$$

(b) The diode size is determined by the ability to match the diode capacitance in the circuit, and the lowest reactive impedance that can be matched is given by

$$\frac{1}{\omega C} = 15 \text{ ohms.}$$

This represents a more realistic practical limit than the 10 ohms suggested by Read.

(c) The maximum d.c. current density is limited by space-charge effects and is given by

$$J_m = \epsilon E_B f_o$$

(d) The correlation between diode breakdown voltage (or depletion-layer width at breakdown) and operating frequency is taken from existing experimental data.

(e) The operating voltage is taken as 20 per cent above the room temperature breakdown voltage to allow for the effects of heating.

(f) The maximum operating temperature is 325°C (that is a temperature rise at the junction of 300°C above ambient).

(g) The d.c.-to-a.c. conversion efficiency is taken from Fig. 5.5.

A more detailed analysis has been presented by Scharfetter [8] considering the limitations imposed by the resistive part of the circuit impedance rather than the reactive component, and including a more realistic model for the thermal impedance of the diode. These and other refinements make Scharfetter's analysis more accurate. However, the simplified treatment presented here retains the essential features and predicts power–frequency limits in reasonable agreement with the more accurate analysis.

The frequency for optimum operation of an IMPATT oscillator is inversely proportional to the depletion-layer width. This has been confirmed by experimental results on both GaAs and Si abrupt-junction devices. The frequency–width correlation from experimental results is, for Si diodes,

$$f = 4{\cdot}2 \times 10^6/w \tag{5.15}$$

and for GaAs diodes.

$$f = 2{\cdot}3 \times 10^6/w, \tag{5.16}$$

where w is in cm and f in Hz. These results are for abrupt-junction diodes, and w represents the total depletion-layer width. We must remember that w does not equal the width of the drift space, since the avalanche zone occupies a significant fraction of the total width. Thus, the proportionality constant between f and $1/w$ is slightly less than $v_s/2$. The smaller magnitude of the proportionality constant for GaAs diodes is not fully explained, but must reflect a smaller saturated electron drift velocity. The drift velocity for electrons in GaAs has not in fact been measured at the large values of electric field present in the avalanche diode. The narrower width of the GaAs diode at a given frequency gives rise to more stringent frequency limitations, as we shall see below.

For abrupt-junction devices the depletion layer width at breakdown is related to the breakdown voltage by $V_B \propto w^{0{\cdot}81}$ This result can be obtained from Figs. 3.7 and 3.8. Thus the operating voltage, which we shall take to be 20 per cent above the room temperature breakdown voltage, is related to the frequency of oscillation by

$$V = 1{\cdot}2\, A_1 f^{-0{\cdot}81} \tag{5.17}$$

for Si diodes and

$$V = 1{\cdot}2\, A_2 f^{-0{\cdot}81} \tag{5.18}$$

for GaAs diodes. The proportionality constants A_1 and A_2 are $1{\cdot}03 \times 10^{10}$ and $6{\cdot}42 \times 10^9$ respectively when V is in volts and f in Hz. The maximum current density based on space-charge effects, $J_m = \epsilon E_B f$, is also frequency-dependent.

Since $E_B = 2V_B/w$ for an abrupt-junction device, the maximum current density is given by

$$J_m = \frac{2\epsilon A_1}{4{\cdot}2 \times 10^6} f^{1{\cdot}19} \qquad (5.19)$$

for Si devices and

$$J_m = \frac{2\epsilon A_2}{2{\cdot}3 \times 10^6} f^{1{\cdot}19} \qquad (5.20)$$

for GaAs. Equations (5.17)–(5.20) can be used to calculate the maximum d.c. power *density* for both materials in the IMPATT mode. The magnitude of the *total* power at any frequency is then obtained, provided we know the maximum diode area that can be used. This is obtained from the impedence limit

$$\frac{1}{\omega C} = \frac{W}{2\pi f\epsilon A} = 15. \qquad (5.21)$$

Substituting for w as a function of f from (5.15) and (5.16), we obtain the maximum diode radius

$$r_m = \left(\frac{4{\cdot}2 \times 10^6}{30\pi^2 \epsilon}\right)^{\frac{1}{2}} \frac{1}{f} \qquad (5.22)$$

for Si and

$$r_m = \left(\frac{2{\cdot}3 \times 10^6}{30\pi^2 \epsilon}\right)^{\frac{1}{2}} \frac{1}{f} \qquad (5.23)$$

for GaAs. The maximum output power for both materials can be evaluated, using eqns (5.17), (5.18), (5.19), (5.20), (5.22), and (5.23), from

$$P_{max} = \eta V J_m \pi r_m^2, \qquad (5.24)$$

where η is the d.c.-to-a.c. conversion efficiency. The results are for Si

$$P_{max} = 2{\cdot}4\,\pi A_1^2 \eta/f^{1{\cdot}62} \qquad (5.25)$$

and for GaAs,

$$P_{max} = 2{\cdot}4\pi A_2^2 \eta/f^{1{\cdot}62} \qquad (5.26)$$

and are shown by the solid lines in Fig. 5.11, using the efficiencies given in Fig. 5.5. The higher efficiency for GaAs is offset by the fact that $A_2 < A_1$. This

is a consequence of the lower proportionality constant appearing in the frequency–width relation discussed above. Notice that the power decreases as $1/f^{1{\cdot}62}$, and not $1/f^2$ as predicted by the simplified transit-time analysis in §5.2 for the idealized Read structure. This difference arises because the breakdown field increases in the abrupt-junction diode as the depletion-layer width decreases.

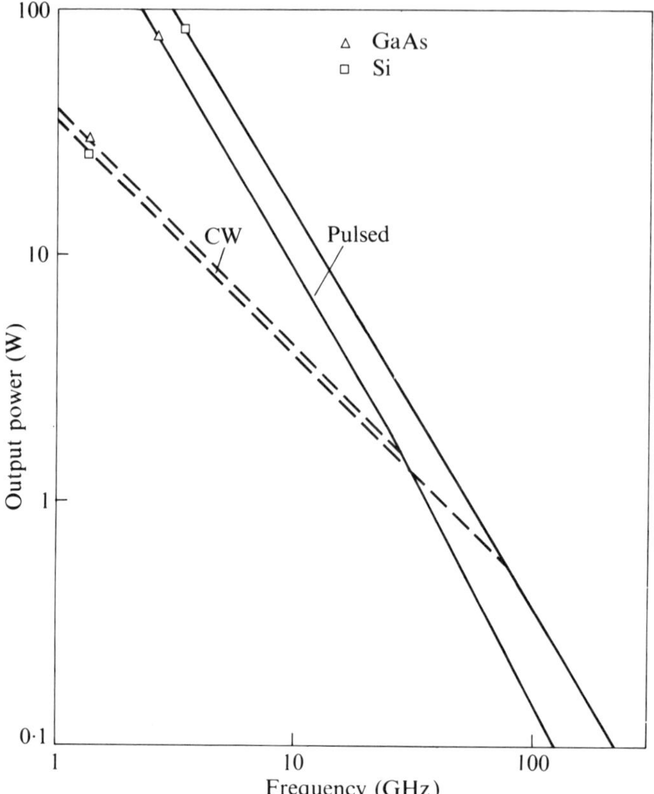

Fig. 5.11 Ouput power versus frequency for Si and GaAs IMPATTs from simplified analysis.

As a result, the operating voltage and maximum current are both somewhat larger than would be predicted using a constant breakdown field. It is well to emphasize that the power–frequency relationship derived here is approximate, and is intended to illustrate the trend rather than predict accurately the absolute power at any frequency. Although the results are undoubtedly accurate to within a factor of two, many uncertainties still remain, particularly in the efficiency fall-off at high frequencies and in the diode width–frequency correlation.

A more rigorous way of calculating the maximum diode area is by using the resistive impedance limitations as considered by Scharfetter [8]. The minimum resistive impedance of the circuit is then determined by the skin depth and is expected to increase with frequency as $f^{0 \cdot 5}$. When this is included in the analysis, the diode area decreases as $1/f^{2 \cdot 5}$ rather than $1/f^2$, and the power decreases as $1/f^{2 \cdot 12}$.

Over a wide range of frequencies the bias current density is restricted by heating effects to a value well below J_m. The thermally limited current density J_{th} is given by

$$P_{th} = VJ_{th}\pi r_m^2 = 300/\theta\,, \tag{5.27}$$

where a maximum temperature rise of 300°C at the junction has been assumed. Using eqns (5.17), (5.18), (5.22), and (5.23), this becomes

$$J_{th} = \frac{300K}{1{\cdot}2A_1}\left(\frac{30\pi^2\epsilon}{4{\cdot}2\times10^6}\right)^{\frac{1}{2}} f^{1{\cdot}81} \tag{5.28}$$

for Si devices and

$$J_{th} = \frac{300K}{1{\cdot}2A_2}\left(\frac{30\pi^2\epsilon}{2{\cdot}3\times10^6}\right)^{\frac{1}{2}} f^{1{\cdot}81}. \tag{5.29}$$

for GaAs. K is the thermal conductivity of the heat sink. For frequencies such that $J_m < J_{th}$ the maximum output power of the oscillator is transit-time-limited. For frequencies such that $J_{th} < J_m$ the power output is thermally limited and is given by

$$P_{th} = \eta VJ_{th}\pi r_m^2. \tag{5.30}$$

It can be shown that in this regime the power decreases as $1/f$. Equation (5.30) has been evaluated for a copper heat sink and the results, are shown for both GaAs and Si by the dotted lines in Fig. 5.11. For frequencies above 90 GHz the Si IMPATT, under CW operation, is transit-time limited. The corresponding frequency for GaAs is about 40 GHz. Under pulsed operation, with short pulses and low-duty cycle such that the junction does not heat up, the transit-time-limited power–frequency relationship applies at all frequencies.

Finally, in this section, we make a comparison of the results obtained here with the more detailed analysis of Scharfetter, referred to earlier. Scharfetter's results for Si diodes are shown in Fig. 5.12. The simplified analysis predicts slightly larger powers. The critical frequencies are almost identical. In Scharfetter's analysis the diode size is scaled by matching the real part of the impedance rather than the imaginary part, and this approach yields a slightly smaller diode diameter.

In Fig. 5.12 the power–frequency relationships, derived by Scharfetter, for Si diodes mounted on diamond heat sinks and Si double-drift diodes on copper heat sinks, are also presented.

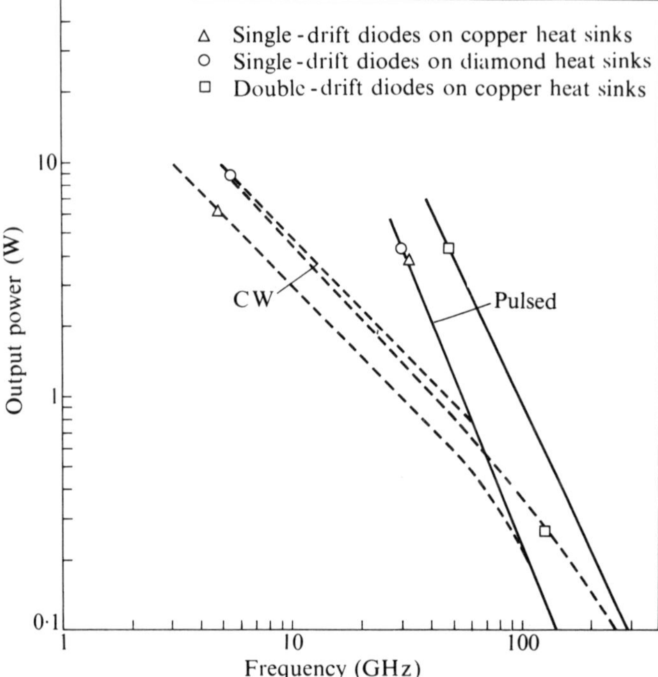

Fig. 5.12 Power–frequency limits for Si IMPATTS; the solid lines represent the transit-time limitations and the dotted lines represent the thermal limitations (after Scharfetter [8]).

References

1. Misawa, T. High-frequency fall-off of IMPATT diode efficiency. *Solid-St. Electron.* **15**, 457 (1972).
2. Read, W. T. A proposed high-frequency negative resistance diode. *Bell Syst. Tech. J.* **37**, 401 (1958).
3. Misawa, T. Minority carrier storage and oscillation efficiency in Read diodes. *Solid-St. Electron.* **13**, 1369 (1970).
4. Kovel, S. R. and Gibbons, G. The effect of unswept epitaxial material on the microwave efficiency and IMPATT diodes. *Proc. IEEE.* **55**, 2066 (1967).
5. Johnson, E. O. Physical limitation on frequency and power parameters of transistors. *IEEE. Intern. Conv. Record,* pt. 5, 27 (1965).
6. DeLoach, B. C., Jr. Recent advances in solid-state microwave generators. *Advances in Microwaves* **2**, Academic Press, New York, p. 43 (1967).
7. Scharfetter, D. L., Evans, W. J., and Johnson, R. L. Double-Drift-region (p^+–pnn^+) avalanche-diode oscillators. *Proc. IEEE.* **58**, 1131 (1970).
8. Scharfetter, D. L. Power–Impedance–frequency limitations of IMPATT oscillators calculated from a scaling approximation. *IEEE Trans. Electron Devices* **ED–18**, 536 (1971).

9. Schroeder, W. E. and Haddad, G. I. Avalanche region width in various structures of IMPATT diodes. *Proc. IEEE.* **59**, 1245 (1971).
10. Udelson, B. J. and Ward, A. L. Computer comparison of n^+pp^+ and p^+nn^+ junction silicon diodes for IMPATT oscillators. *Electron. Lett.* 7, 723 (1971).
11. Ying, R. S., X Band n^+np^+ diode. *Electron. Lett.* 8, 297 (1972).
12. Carslaw, H. S. and Jaeger, J. C. *Conduction of Heat in Solids.* Clarendon Press, Oxford (1959).

6 Practical IMPATT oscillators

6.1 Device fabrication

Most high-frequency semiconductor devices, IMPATT diodes included, are fabricated using epitaxial material which consists of a thin high-resistivity layer on a thick low-resistivity substrate. The epitaxial layer is grown by a chemical vapour-deposition technique at elevated temperatures, and the properties of the layer are controlled by adjusting the growth conditions, such as the temperature of the substrate and the chemical composition of the gas. The epitaxial layer constitutes the active region of the device, the substrate serving primarily as a mechanical support for the layer. The main property required of the substrate is that it be heavily doped to minimize the parasitic and contact resistances. The properties of the device are determined by the doping density and profile and the thickness of the high-resistivity layer. A steep impurity concentration gradient exists at the layer–substrate interface and may lead to contamination of the epitaxial layer by the out-diffusion of impurities from the heavily doped substrate. This may occur either during the epitaxial growth itself or during subsequent high-temperature operations on the slice in the device fabrication process. This effect, which is particularly important for high-frequency devices, has to be minimized, otherwise control of the layer properties will be lost. Thus, the impurity used to dope the substrate must be a slow diffusing species, and in practice this frequently specifies the polarity of the device. The substrate impurity must also be one which has a high solid solubility in the semiconductor, in order to obtain the high doping levels and low resistivity required.

In Ge, since donors diffuse more rapidly than acceptors [1], p-type expitaxial layers on p^+ substrates can be grown with greater control than n-type layers on n^+ substrates. The Ge substrate is usually doped with Ga, which has a solid solubility [2] of about 10^{20} atoms cm^{-3} and produces a substrate resistivity of about 0·0006 Ω cm. The n^+–p junction is formed in Ge by diffusing donors, As or Sb, into the surface of the p-type epitaxial layer to a depth of about 1 μm. Arsenic is one to two orders of magnitude more soluble than Sb [2] and therefore has the advantage of producing high surface impurity concentrations. This facilitates the fabrication of low-resistance contacts to the diffused layer.

In Si, where acceptors in general diffuse more rapidly than donors [1], the opposite polarity is generally used. Sb, which is a slow diffuser in Si, is often used to dope the substrate, but is somewhat limited in its solid solubility, the lowest resistivity attainable being about 0·01 Ω cm. This resistivity is adequate for many applications, particularly at the lower frequencies. At high frequencies a lower resistivity is desirable since skin effects and the small diode size combine to make the parasitic resistance more troublesome, and As-doped substrates with a resistivity of about 0·001 Ω cm are generally employed. In Si, the junction is formed by diffusing B into the n-type epitaxial layer to a depth of about 1 μm from the surface. An alternative technique for constructing an abrupt-junction Si IMPATT is by the fabrication of a Schottky barrier junction [3]. This type of junction, as an IMPATT oscillator, possesses essentially identical properties to those of the abrupt p–n junction and is constructed by depositing a metal layer, either by vacuum evaporation, sputtering, or plating, on the surface of the n-type semiconductor.

The only other semiconducting material in which IMPATTs have been constructed is GaAs. GaAs epitaxial materials technology has been actively developed over the past five years for the development of the Gunn-diode oscillator. For this application n-type epitaxial material is always used since the bulk negative resistance of GaAs is found only in n-type samples, and as a result the GaAs materials technology has been concentrated on this resistivity type. The junction in GaAs is formed either by the Schottky barrier technique or by diffusing Zn acceptors into the n-type epitaxial layer. In practice, because of the ease of fabricating Schottky barriers on n-type GaAs, and because of the less advanced diffusion technology by comparison with Si, the Schottky barrier technique has found widespread application in the fabrication of GaAs IMPATTs. The high electron mobility makes the doping level of the substrate not as critical as in Si or Ge, and substrate resistivities of 0·001 Ω cm are readily obtained using Te, Sn, or Si as the dopants.

For all three semiconductors the doping level and thickness of the epitaxial layer are chosen so that the depletion-layer edge at breakdown sweeps almost to the epitaxial layer–substrate interface. This is required to ensure that the parasitic resistance contributed by any undepleted high-resistivity layer is small. On the other hand, 'punch-through' of the depletion layer to the heavily doped substrate must be avoided, as discussed in §4.7, in order to preserve a field profile in the diode which is favourable to the production of high-efficiency IMPATT oscillations. The doping level in the epitaxial layer determines the active region width and therefore the frequency of oscillation.

Diode construction techniques

Several techniques have been used for constructing IMPATT diodes. A necessary property of any such technique is that it should produce a structure from which heat can be removed efficiently. This means that the active region of the diode, where the heat is generated, must be located close to a large, high-thermal-conductivity heat sink. A batch fabrication process, developed by Zettler and

Cowley [4], in which a heat sink is formed on the semiconductor slice and forms an integral part of the diode chip, has been used in many laboratories. This fabrication process, called the integral heat sink technique, is illustrated in Fig. 6.1 for the construction of a Si diffused-junction IMPATT. Following the formation of the junction by diffusing boron into the n-type epitaxial layer, a metal contact is applied to the p^+ surface by vacuum evaporation. The contacts are described in more detail in the next section. A thick metal layer is subsequently electroplated onto the evaporated contact to form the integral heat sink. This metal has a high thermal conductivity, usually Au, Ag, or Cu, and the thickness of the electroplated layer is about 100 μm. Using the electroplated metal layer as a support for the semiconductor slice, the n^+ substrate is thinned to a thickness of approximately 20 μm either by mechanical polishing or chemical etching. An evaporated metal contact is then deposited on the substrate side of the slice and photoengraving and etching techniques are used to define an array of metal dots on the back surface and to etch out the devices.

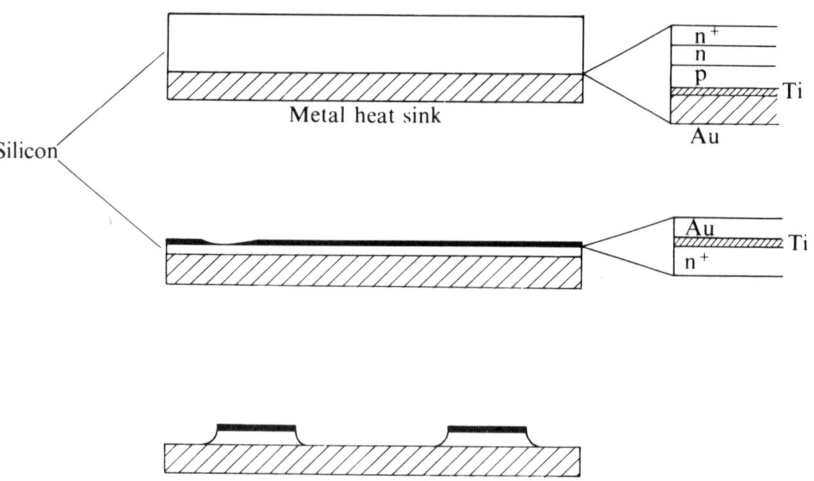

Fig. 6.1 The integral heat sink technique for the fabrication of IMPATT diodes.

At this stage, an array of mesa diodes is formed on the electroplated metal heat sink with the active region of each diode located about one μm from the heat sink. The metal is finally cut up into square dice typically 500 μm square, each one supporting an individual diode which, at X band, is typically 100 μm to 150 μm in diameter. These device chips are then bonded into a microwave package illustrated in Fig. 6.2. The base of the package is constructed of oxygen-free high-conductivity copper which is plated with a thin layer of gold to prevent oxidation of the copper during the bonding operation. The bonding is often carried out by using a low-melting-point alloy, such as Au–Sn, which alloys with the metal heat sink on the device chip and with the gold-plated layer on the

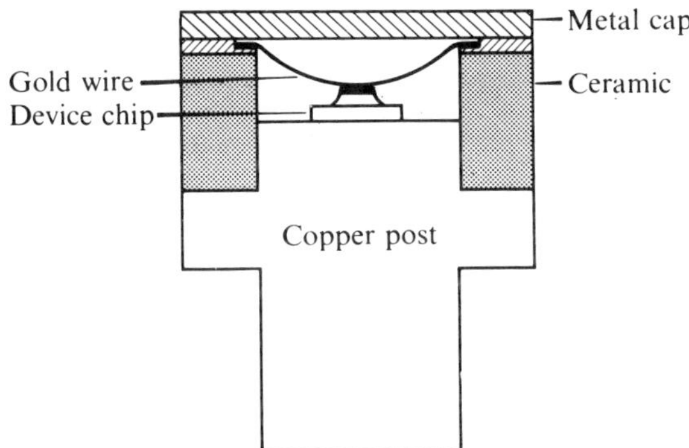

Fig. 6.2 Packaged device.

package post. However, the thermal conductivity of the alloy is usually low and the uniformity of the device–package interface is often quite variable. An alternative method is an ultrasonic bonding technique, in which the metal heat sink is bonded directly to the gold-plated package post using a combination of heat, ultrasonic energy, and pressure. This method is cleaner than the alloy technique and usually produces better and more consistent high-quality bonds.

Several special techniques have been developed for constructing high-power devices. We have discussed one of these, diamond heat sinking, in the previous chapter. Improved thermal impedance and power capability have also been obtained using arrays of small diodes and by employing ring and stripe geometries in place of the conventional solid circular structure.

Since the diode area is limited by the impedance of the circuit, no increase in the total active area is obtained using a parallel array of devices. However if the individual chips in the array are small, a large number n can be accommodated and the thermal impedance of the array decreases as $n^{\frac{1}{2}}$, compared with a large single diode of the same total area. A larger active area is possible only if a series or series–parallel array is constructed, and this presents severe practical problems, since efficient heat removal from a series stack of devices cannot be achieved using conventional techniques. A novel solution to this problem has been devised by Josenhans, [5] utilizing the insulating properties of diamond in addition to its high thermal conductivity. By mounting the diode chips on a selectively metallized type IIa diamond as shown in Fig. 6.3, the chips are connected in series electrically but in parallel thermally. In this way good heat sinking is provided equally for all the chips while, at the same time, the electrical impedance problems of the parallel array are avoided. The second approach to increase the power capability of IMPATTs involves altering the diode geometry to achieve a smaller thermal impedance [6]. This method can, of course, be used in conjunction with the above techniques. In a ring structure the hot central portion of the solid circular diode is removed, and more efficient spreading of the heat flux results. A similar improvement can be realized with

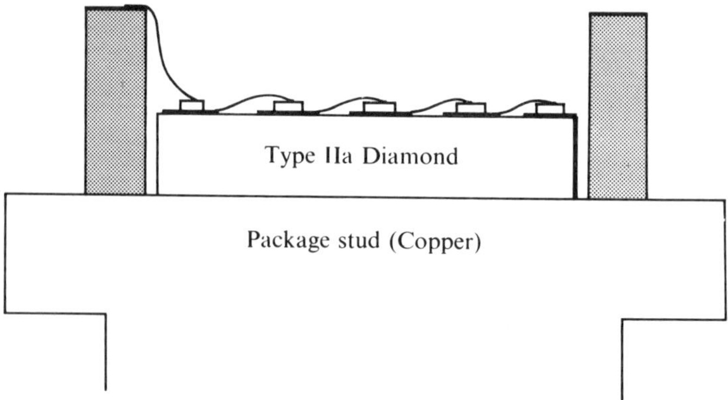

Fig. 6.3. Series combination of IMPATT diodes on diamond heat sink (after Josenhans [5]).

a stripe geometry. In both cases, the thermal impedance decreases as the width decreases and the length increases. The limit is set by the practical problems inherent in fabricating long narrow devices. In practice, an improvement of about a factor of two over the equivalent area solid circular structure can be obtained.

Thermal Impedance

The thermal impedance of an avalanche diode can be measured non-destructively from the d.c. reverse voltage–current characteristic, provided the space-charge resistance and temperature coefficient of the breakdown voltage of the diode are known. The effects of temperature on the diode breakdown characteristics have been discussed briefly in §3.5. The thermal impedance θ of the diode is obtained from equations (3.29) and (3.32), namely

$$\theta = \frac{1}{\beta_0 V V_B(T_0)} \left[\frac{V - V_B(T_0)}{I} - R_{sc} \right], \tag{6.1}$$

where V and I are the operating voltage and current bias. The space-charge resistance is first obtained by measuring the d.c. current–voltage characteristic using pulsed bias. Provided the pulse length is very much shorter than the thermal time constant of the diode, heating effects during the pulse are negligible and the space-charge resistance is simply the slope of the curve in breakdown. The temperature coefficient of the breakdown voltage is measured by d.c.-biasing the diode into breakdown, and measuring the rate of change of the voltage with change in ambient temperature at low current levels.

An alternative technique is to measure the a.c. impedance of the diode biased well into breakdown [7]. By measuring the impedance as a function of frequency, the space-charge resistance term can be separated from the effects of heating. At low frequencies the measured resistance is the combined effect of heating and the space-charge resistance, while at high frequencies, if the period of the a.c. signal is smaller than the thermal time constant of the diode, the measured value is simply the space-charge resistance.

Contacts

The interface between a semiconductor and a metal contact is quite complicated. In almost all semiconductor devices it is necessary to provide ohmic contacts through which the current enters and leaves the device. By definition an ohmic contact is one in which the magnitude of the potential barrier at the metal–semiconductor interface is independent of the magnitude and direction of the current flow through the contact. Two types of contact exist, depending on the metal–semiconductor work function. If the metal work function ϕ_m is larger than the electron affinity ϕ_s of the semiconductor, the barrier behaves like a p–n junction and is a rectifying contact. However, if the metal work function is smaller than the electron affinity of the semiconductor, the contact potential $(\phi_m - \phi_s)$ is of the opposite polarity and is a non-rectifying or ohmic contact. These contacts are characterized by the existence in the semiconductor of a depletion layer or an accumulation layer respectively, as shown in Fig. 6.4. The

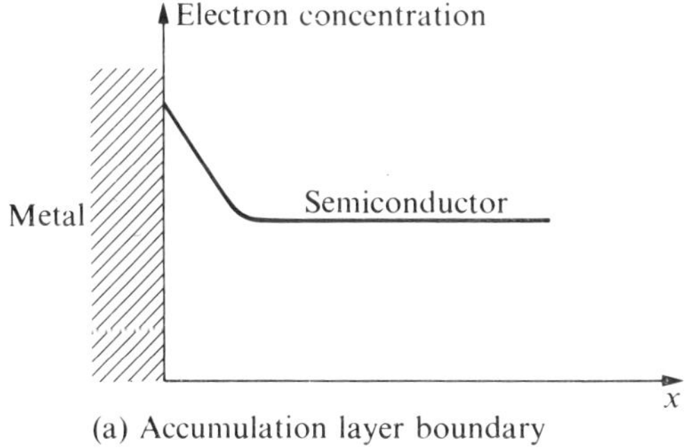

(a) Accumulation layer boundary

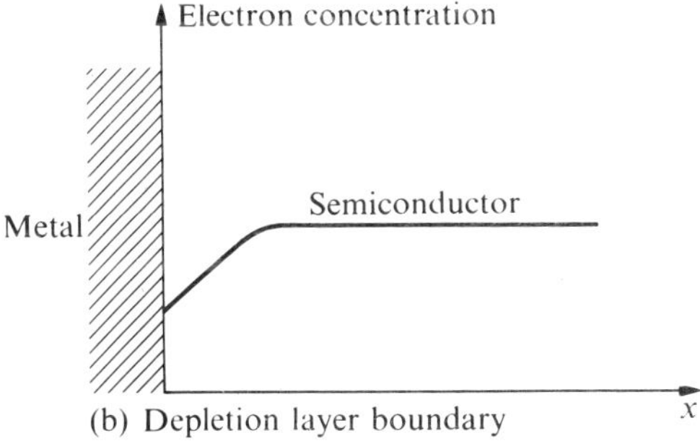

(b) Depletion layer boundary

Fig. 6.4 Electron concentration at metal–semiconductor boundary.

electron concentration n_0 at the surface of the semiconductor under the metal contact is determined by the metal–semiconductor work function ϕ_{ms},

$$n_0 = N_c \exp\left(-\frac{\phi_{ms}}{kT}\right), \tag{6.2}$$

where N_c is the effective density of states in the conduction band of the semiconductor. When the metal–semiconductor work function is positive, a depletion layer is present at the boundary ($n_0 < n$), and when ϕ_{ms} is negative the electron concentration at the surface is enhanced ($n_0 > n$) and the contact is characterized by a charge accumulation layer at the boundary. In the accumulation layer contact, a large current can flow with little change in the barrier potential. This latter type of contact is ohmic. In most practical cases, however, the potential barrier at the interface is not dependent on the work function of the metal. This arises because of the high density of surface states on the semiconductor, which pins the Fermi level at an energy about $E_g/3$ above the valence band. This is the case in covalent semiconductors such as Ge, Si, and GaAs, and the resulting barrier at the interface is rectifying with an almost constant value of $2/3\ E_g$ for all metals. However, the resistance of the depletion layer contact can be reduced to a very small value by using a heavily doped semiconductor region adjacent to the metal. If the doping level in the semiconductor is sufficiently high, the depletion layer is very thin and current passes between the metal and semiconductor by quantum mechanical tunnelling through the potential barrier. This type of contact, although not strictly ohmic, can be made with interfacial resistances of $\sim 10^{-6}\ \Omega\text{cm}^2$, and its use has become almost standard practice. When applied to a high-resistivity semiconductor, it is necessary first to form a heavily doped layer at the surface adjacent to the metal. This is usually accomplished by diffusion, epitaxial growth, or alloying. An alloyed 'ohmic contact' is formed by melting the semiconductor surface in contact with a metal containing a small percentage of dopant. The metal either has a low melting temperature or is one which forms a eutectic with the semiconductor at a low temperature. When the melt is cooled, a heavily doped region is formed on the semiconductor surface below the metal. The metal–semiconductor contact is then at the interface between the metal and the heavily doped re-grown region. Alloyed contacts are not particularly reproducible and are usually avoided wherever possible. Greater control and reproducibility are obtained with both epitaxy and diffusion, and these processes are generally preferred.

Two important properties of any contact to a semiconductor device, in addition to low resistance, are that it adhere well to the surface of the semiconductor and that it be stable when subjected to high temperatures and high humidity. These factors are particularly important for high-power devices such as avalanche diodes. It is difficult to find one metal which simultaneously satisfies all of these requirements, and as a result contacts are most often layered structures consisting of more than one metal. Gold is almost always used as the top layer in such a contact, since it does not easily oxidize and provides a surface to which it is easy to bond a wire. Since a very thin layer of oxide is always present on the semiconductor surface, the metal deposited on the surface must adhere to this oxide. Metals such as Ti or Cr, which oxidize more easily than the semiconductor, will reduce the oxide and adhere well to the surface. Thus composite contacts of Ti–Au or Cr–Au are often used. However, the stability of the Ti–Au contact is unsatisfactory, since Ti and Au interact at temperatures of 250°C to 300°C. A more stable metallurgical system is obtained by incorporating a barrier metal between the Ti and Au which prevents this interaction from taking place. The metal most useful for this purpose is Pt which is quite impervious to the penetration of Au. In fact Pt can also be used for the contact layer adjacent to the semiconductor surface in place of Ti provided that it is deposited by sputtering in order to get adequate adherence to the surface. Pt has an additional advantage when used on Si devices. When Pt in contact with Si is heated to around 700°C, PtSi is formed by a solid–solid reaction which does not involve the formation of a liquid alloy. PtSi is an extremely stable compound and is the basis of the contact structure used in passivated-beam lead devices [8] but can, of course, be used on any Si device in which a reliable contact is required. Contact to the PtSi layer is made by a layer of Ti or Cr followed by a Pt barrier layer and by the top Au contact. The PtSi contact can be used to form the active junction of a Si Schottky barrier IMPATT [3] or a low-resistance contact to a diffused-junction device, depending on the doping level of the Si underneath the PtSi.

The contact applied to the diffused side of the p–n junction IMPATT diode is close to the active region, and can have a deleterious effect on the performance of the device if this contact injects minority carriers into the semiconductor. If these minority carriers diffuse to the active region, the conversion efficiency and noise performance of the oscillator may be severely degraded. From measurements on Ge IMPATTs, Decker *et al.* [9] have shown that if minority-carrier injection constitutes as little as 6 per cent of the total contact current, the oscillator efficiency is reduced by one order of magnitude and the FM noise increased by a factor of two. The injected minority-carrier current modifies the phase relationship between the a.c. current and voltage in the diode, and reduces the negative resistance. Furthermore, when minority-carrier injection is present the negative resistance of the diode decreases more rapidly with increasing a.c. voltage swing, producing a situation which is not favourable to high-efficiency and high-power operation of the oscillator.

Thus, as for almost every semiconductor device, the contacts play a major role, affecting both the electrical performance and reliability of the device.

Special fabrication techniques for high-frequency diodes

At frequencies above about 50 GHz, several new techniques are incorporated into the process for fabrication of IMPATT diodes. These are required because of the small size of the devices and because, in general, parasitic effects become more troublesome at high frequencies. To match the circuit impedance at high frequencies, the diode diameter is restricted to approximately 20 μm at 100 GHz. Using the integral heat sink fabrication process described above, in which the device chips are isolated by etching through the semiconductor from the back of the slice, the substrate must be thinned to about 5–6 μm in order to control the junction dimensions without severe under-etching of the back contacts. The thinning of the substrate is also desirable to minimize the parasitic resistance of the diode, since the skin depth in the substrate at 100 GHz is about 5 μm. The packaging techniques used at low frequencies are quite unsatisfactory at frequencies much above 30 GHz, since the parasitic capacitance and inductance of commercially available packages give rise to unwanted resonances. To overcome this problem a new packaging technique [10] has been developed which produces much smaller parasitic capacitance and inductance. This technique, illustrated in Fig. 6.5, involves bonding metallized quartz stand-offs on either side of the diode chip, which is bonded to a gold-plated copper stud. The back contact to the diode is made by thermocompression bonding a low-inductance gold tape between the stand-offs and the back metallization on the chip. The overall dimensions of the device are quite small and depend on the intended

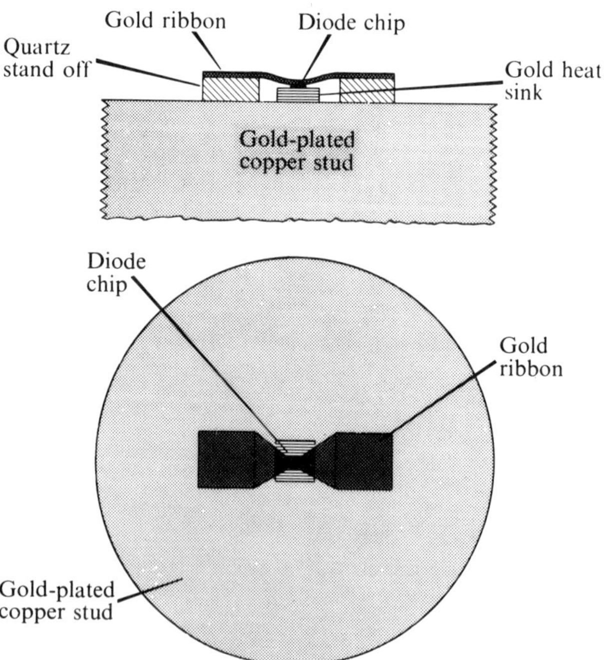

Fig. 6.5 Millimetre-wave IMPATT-diode packaging technique.

frequency of operation. For frequencies in the range of 50 to 75 GHz, the stand-offs are typically 100 μm square 50–75 μm high, and are placed approximately 200 μm apart.

The diode active region is less than 0·5 μm wide at frequencies above 50 GHz, and the growth of suitable epitaxial layers requires a degree of control close to the limits of present-day technology. Only by growing at low temperatures in fast response reactors can the thickness and doping levels be obtained within reasonable tolerances.

The best IMPATT oscillator structure at high frequencies is the double-drift p^+–p–n–n^+ device [11]. Although this structure could be fabricated, using epitaxial techniques, by successively growing n and p layers on an n^+ substrate following by a p^+ diffusion, the experimental devices constructed to date [12] in Si have all been fabricated using a combination of epitaxial growth and ion implantation. The ion implantation technique offers several advantages when a thin precisely controlled doped region is required, since the process is carried out at low temperatures. The doped region is formed by bombarding the semiconductor sample with very energetic (25 to 300 keV) impurity ions, which penetrate into the sample to a depth determined by the energy. The distribution of impurities in the implanted layer is a Gaussian distribution about the mean range, and the impurity density is determined by the particle flux. The impurity distribution in an ion-implanted double-drift structure is shown in Fig. 6.6 [12]. This distribution is obtained by multiple implants of boron into an n-type epitaxial layer on an n^+ substrate. Three implants at different energies are used to form the p-region and the boron flux is chosen so that the acceptor concentration is exactly twice the donor level in the epitaxial layer. In this way the net acceptor level ($N_A - N_D$) in the p region is exactly equal to the net donor level in the n region. The p^+ surface layer which is required for contacting purposes may be formed either by a shallow boron diffusion or by a low energy high-dose boron implantation. Although the bombardment produces a large amount of damage in the crystal lattice this is later removed by annealing at temperatures in the range 700–900°C and high-quality, defect–free junctions are produced.

6.2 IMPATT oscillator circuits

Practical circuits and stability

To construct an oscillator, the diode is embedded in a microwave circuit and connected to a load through a matching network. In practice the circuit is either a coaxial line or a section of waveguide. In the coaxial system the diode is mounted at the end of the line, and the tuning and matching is accomplished by means of small movable sections of low-impedance line inserted in front of the diode. In the waveguide structure, the diode is mounted across the guide,

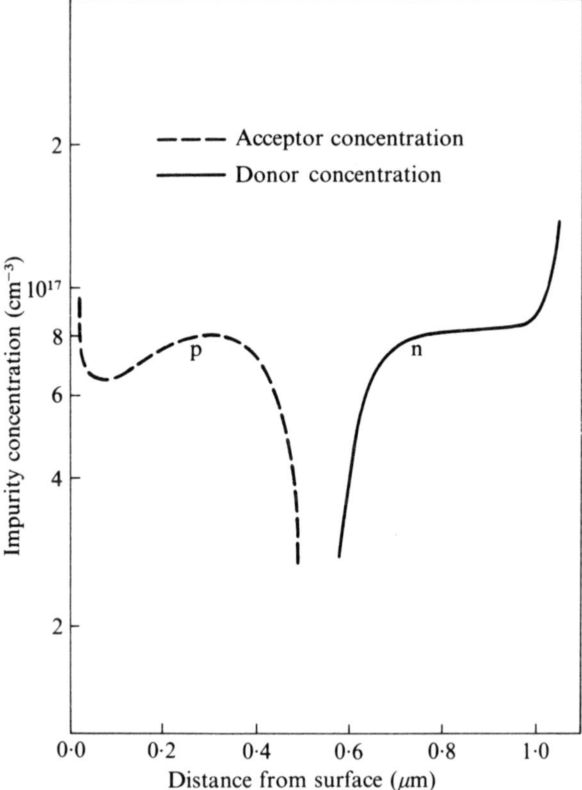

Fig. 6.6 Impurity distribution in ion-implanted double-drift diode.

and tuning and matching are provided by a movable short-circuit behind the diode and a matching element between the diode and the load. The simplest equivalent circuit for the IMPATT diode is a negative conductance G_D in parallel with a capacitive susceptance B_D, and the equivalent parallel circuit of the cavity, as seen at the diode terminals, is represented by a positive conductance G_L in parallel with an inductive susceptance B_L. The condition for a stable oscillator is

$$|G_D| = |G_L|,$$
$$|B_D| = |B_L|. \tag{6.3}$$

The stability of the oscillator is ensured by the net zero conductance of the circuit–diode combination, and the frequency of oscillation is determined by the resonant frequency of the system, where the total susceptance is zero. In general the two conditions of equation (6.3) will not be satisfied simultaneously, since the negative conductance of the diode will not in general be equal to the load conductance at the resonant frequency. However, provided the magnitude of G_D is larger than G_L, oscillations will build up. The oscillation is started by a random noise fluctuation which will grow only if the total conductance of the system is negative. For stability, it is a required property of any oscillator of this type, that as the oscillation amplitude grows, the magnitude of the negative conductance of the active element decreases. This has been shown to be the case, in fact, for the IMPATT oscillator by Scharfetter and Gummel [13], who have also shown that the increasing voltage swing leaves the susceptance of the diode relatively unchanged, and therefore the frequency of oscillation remains fixed as the amplitude builds up. The oscillation continues to build up for several cycles until the magnitude of the diode negative conductance decreases to the value of the positive conductance of the load. The oscillator is then stable. This is illustrated in Fig. 6.7 by means of a set of large-signal admittance plots

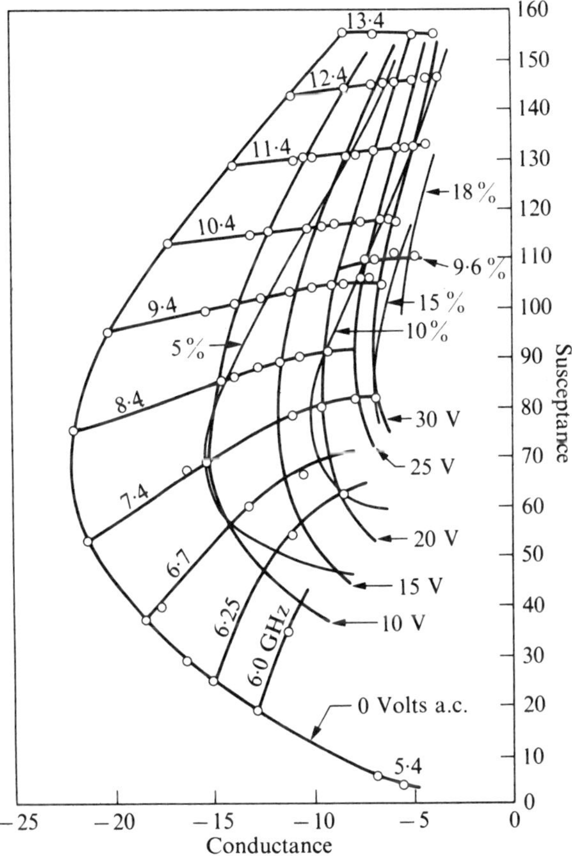

Fig. 6.7 Diode admittance (susceptance versus conductance) as a function of frequency and a.c. voltage amplitude, and resultant efficiency indicated. Current density: 200 A/cm^2 (after Scharfetter and Gummel [13]).

for a Si Read diode computed by Scharfetter and Gummel [13]. In this figure, G_D versus B_D is shown as a function of frequency for different values of voltage swing.

An experimental coaxial circuit [14] for testing high-efficiency IMPATT oscillators is shown in detail in Fig. 6.8. This type of circuit is commonly used up to frequencies of about 12 GHz. Although coaxial circuits can be used at higher frequencies, they must be designed to suppress all propagating modes other than the TEM mode, and in practice waveguide circuits are more frequently used for frequencies above X band. To remove heat from the device, the diode is always inserted through a relatively massive heat sink usually made of copper. In the circuit shown in Fig. 6.8, d.c. bias is fed in to the diode via a bias tee, which presents an infinite d.c. resistance looking towards the load and a high a.c. impedance looking back into the d.c. supply. This ensures that the d.c. voltage appears only across the diode, and prevents leakage of the microwave power out of the bias lead. The microwave power

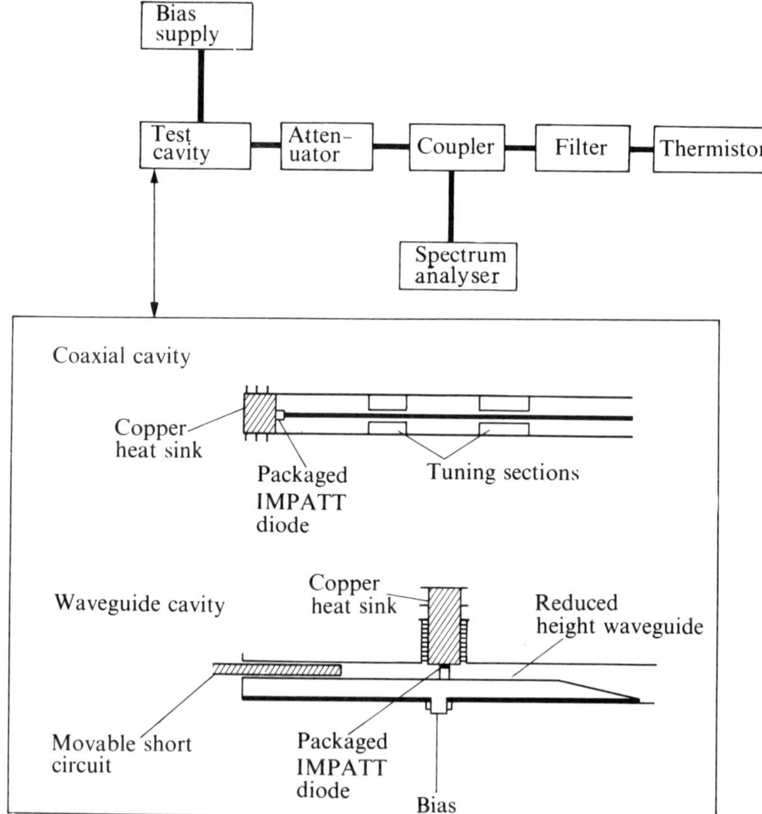

Fig. 6.8 Oscillator test circuit: coaxial and reduced-height waveguide cavities.

passes through a 10-dB directional coupler in such a way that 10 per cent of the power is coupled out to a spectrum analyser which may be used to measure the frequency, noise, and stability of the signal. 90 per cent of the power passes straight through the coupler and into a detector or power-measuring device such as a matched thermistor or calorimeter. Several attenuators are incorporated into the power-measuring branch to prevent overloading the power-measuring device, and filters are often incorporated to eliminate spurious signals far from the frequency of oscillation.

A typical reduced-height waveguide cavity is also illustrated in Fig. 6.8. Both full-height and reduced-height cavities are used. The waveguide impedance level is lowered by reducing the waveguide height, and since the impedance of the diode is always very much smaller than the transmission system to which it is matched, the reduced height guide is a very convenient structure for testing IMPATT oscillators.

At frequencies where both coaxial and waveguide cavities have been used, comparable oscillator efficiency has been obtained from either circuit configuration. However, many of the other properties of the device such as noise, stability, and tunability are more significantly affected by the cavity type. The waveguide cavity possesses a higher Q factor than a coaxial system. This leads to greater stability and lower FM noise in waveguide circuits. However, the higher Q limits the ability to electronically tune or frequency-modulate the oscillator. The noise level of an IMPATT oscillator is determined by statistical fluctuations inherent in the avalanche multiplication process. Under large-signal oscillator conditions, the principal factor affecting the noise is jitter from cycle to cycle in the current pulse produced in the avalanche zone. Since the avalanche current builds up in each cycle rapidly from a very low level, small fluctuations in the carrier density when the current begins to build up will produce significant phase fluctuations of the current pulse from cycle to cycle. This effect produces fluctuations in the frequency and amplitude of the signal. The FM noise produced by this mechanism presents problems when an IMPATT oscillator is used in certain applications. The noise performance of an IMPATT diode is characterized by a noise temperature T, and the FM noise is expressed as a frequency deviation Δf, given by

$$\Delta f = \frac{f_0}{Q}\sqrt{\left(\frac{kTB}{P}\right)}, \tag{6.4}$$

where f_0 is the oscillator frequency, Q is the loaded Q of the circuit, P is the output power, and B is the bandwidth. The oscillator can be stabilized using a high Q cavity, and the FM noise can be significantly reduced in this way. However, in many applications it is required to frequency-modulate the oscillator by electronic means and a high frequency-modulation-sensitivity is not compatible with a high Q. Frequency modulation can be performed in several ways, the simplest of which is by variation of the bias current in the diode. The frequency of oscillation is dependent on the current through the diode, and by applying a low-frequency signal to the bias current, the oscillator frequency can be modulated. An alternative modulation technique is by electronic variation of the circuit susceptance. This can be performed by incorporating a reverse-biased varactor diode in the cavity. The capacitance of the varactor, which is voltage-dependent, contributes to the reactance of the cavity as seen at the terminals of the IMPATT diode. By applying a modulating voltage to the varactor, the resonant frequency of the circuit is varied and the frequency of the IMPATT oscillator is modulated. The frequency-modulation-sensitivity decreases as the Q of the circuit is raised. Thus, in applications where both low FM noise and high frequency-modulation-sensitivity S are required, the diode possesses a figure of merit $S/\Delta f$ which is independent of the circuit Q and depends only on the properties of the diode [15].

A large-signal theory for the noise performance of the IMPATT oscillator has not yet been developed. Owing to the complexity of the problem, an analytic solution is not possible without introducing unrealistic simplifying assumptions. On the other hand, a numerical solution of the problem is time-consuming and expensive. Small-signal noise theory [16] predicts that the noise level of the device decreases when operated above the frequency at which optimum efficiency is obtained. Thus, there is expected to be a trade-off between high efficiency and high power on the one hand and low noise and high frequency-modulation-sensitivity on the other. For any particular application the operating conditions must be chosen accordingly.

Circuits for millimetre-wave IMPATTs

When the wavelength is small ($<$ 1 cm), circuit design in general becomes more difficult, and higher-precision techniques are required for constructing the cavities. Greater care must be taken at discontinuities which may be present in the neighbourhood of the diode associated with the mounting stud and bias post. These difficulties are compounded by the small diode size and specialized packaging techniques used at high frequencies. The skin depth in most metals is less than 1 μm at 50 GHz and the losses can be quite large in the walls of the cavity unless care is taken to produce a smooth finish. Thus it is usual to polish and gold-plate these surfaces, especially in the neighbourhood of the diode.

A resonant-cap configuration, shown in Fig. 6.9, has been used successfully for frequencies up to 100 GHz and above [17]. The cap is a circular metal disc attached to the end of the bias post which is inserted through the top wall of the waveguide. The cap forms a resonant radial cavity around the diode and determines the frequency of the oscillator, the diameter of the cap being equal to $\lambda_0/2$. The sliding short-circuit and E–H tuner provide adjustments to optimize the matching of the diode resistance to the load resistance. The frequency of the oscillator is not easily tuned in this type of configuration, since the dimensions of the radial cavity set the frequency of oscillation quite precisely and the bandwidth of the cavity is very narrow, typically one or two per cent. The frequency can be altered by changing the cap size. This type of cavity is quite useful for the experimental characterization of IMPATT diodes in the laboratory but is quite limited in its applications.

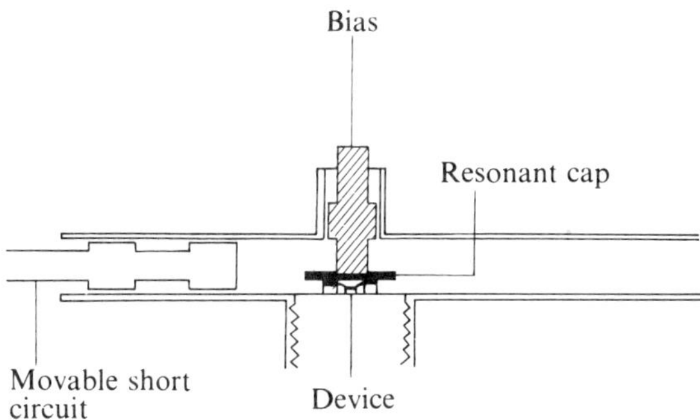

Fig. 6.9 Resonant cap cavity for high-frequency IMPATT oscillators.

A novel circuit design for IMPATT oscillators has been developed by Kurakawa and Megalhaes [18] and applied successfully to millimetre-wave devices by Kenyon [19]. This circuit is a coax-waveguide configuration and is shown schematically in Fig. 6.10 along with its idealized equivalent-circuit representation. The diode is mounted at the end of a coaxial transmission line which is terminated at the other end by a matched load. A low-impedance matching section, $\lambda/4$ long, is inserted adjacent to the diode. The coaxial line is series-loaded by a high Q waveguide cavity at a distance l from the diode as shown. In this circuit the sliding short is adjusted to optimize coupling from the coaxial line to the waveguide cavity, and the coupling between the cavity and the waveguide system is adjusted by rotation of the cavity so as to cross the waveguide planes. In this way a high-frequency, high-Q resonant cavity can be constructed in a configuration which is amenable to analysis.

6.3 Performance of IMPATT oscillators

Power, efficiency, and frequency

The state of the art in IMPATT oscillator performance has changed rapidly over the past five or six years. The development of the oscillator has been directed largely towards the improvement of power and efficiency under CW conditions. After the realization that the complicated Read structure was not essential and that useful powers and efficiencies could be obtained from the simple p^+–n (or n^+–p) abrupt-junction diode, progress was made quite rapidly. The most

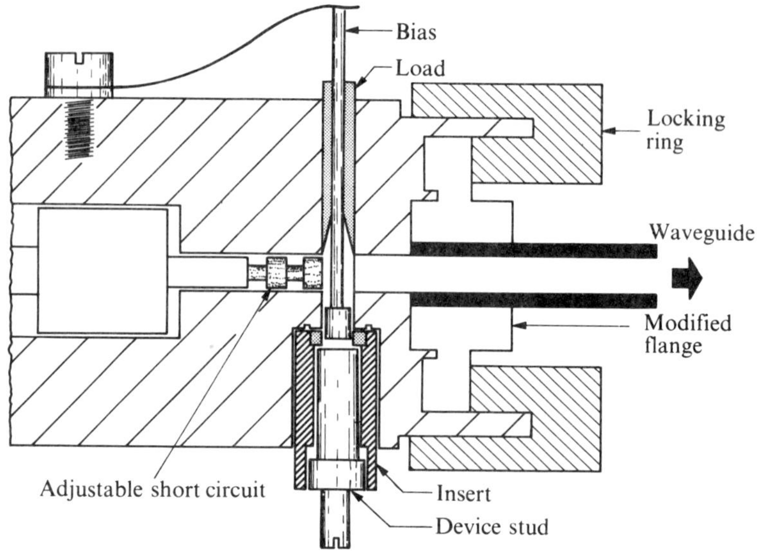

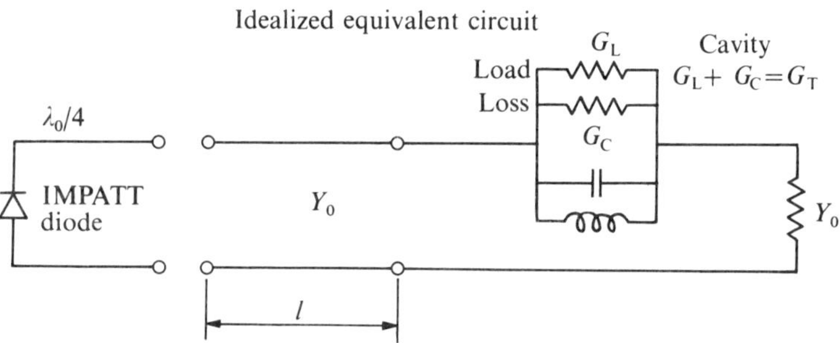

Fig. 6.10 Coax-waveguide circuit for millimetre-wave IMPATTs (after Gibbons *et al.* [31]).

significant technological developments which led to improved oscillator power and efficiency were

(a) the use of flip-chip or junction-side-down mounting for improved heat dissipation and high-power capability,

(b) the realization of the severe degradation of efficiency produced by one or two ohms of positive parasitic resistance and of the necessity of eliminating any excess epitaxial material,

(c) the use of diamond heat sinks, and arrays of diodes to decrease thermal impedance and increase output power,

(d) the demonstration of improved efficiency using GaAs rather than Si to exploit the equal ionization rates and high electron mobility of that material, and

(e) the development of a reliable circuit technique for combining the output powers of several separately packaged devices [20, 21].

The first CW IMPATT oscillator producing 1 watt of output power was constructed by Misawa [22] at Bell Telephone Laboratories in 1967 using a Si p^+–n abrupt-junction device mounted junction-side-down on a copper heat sink. The efficiency of this oscillator was eight per cent. Using a similar device mounted on a diamond heat sink, Swan [23] obtained an output power of 2 watts at 12 GHz, and by constructing arrays of up to five diode chips in parallel in one package the same team of workers at Bell Telephone Laboratories obtained powers of 5 watts CW on a diamond heat sink and 3 watts CW on a copper heat sink. Using the series mounting techniques on a selectively metallized diamond heat sink (Fig. 6.3), Josenhans [5] obtained 5 watts CW at 12 GHz from a five-diode array. These results represent the state of the art in 1972 for X band p^+–n Si IMPATTs, and no significant improvements have been realized in the last few years. Furthermore, no significant improvements in the power and efficiency are to be expected from this structure unless some unforeseen breakthrough in IMPATT design or device technology occurs. However, recent work on n^+–p devices [24, 25] shows promise of giving rise to a new breed of Si IMPATTs with improved efficiency, as discussed in §5.1. The research and development effort in Si IMPATTs in the last few years has been concentrated on high-frequency (50–150 GHz) devices. The oscillator design at these frequencies was again pioneered at Bell Telephone Laboratories using both single-drift and double-drift structures. The first CW oscillators in the 50–100 GHz frequency range were reported by Misawa [17] in 1968. Using single-drift diodes Misawa obtained CW output powers of about 100 mW at frequencies from 50 to 80 GHz with conversion efficiencies of 3 per cent. Subsequent improvements in output power and efficiency were steadily made by the development of a more sophisticated device technology incorporating the ion-implanted double-drift structure. The improved technology has led to the production of double-drift Si IMPATTs producing 1 watt of CW output power at 50 GHz with 14 per cent efficiency, and 250 mW of CW output at 100 GHz with 8 per cent efficiency. These results [12, 30] reported in 1971 undoubtedly represent performance levels close to the attainable limits for single-chip Si IMPATTs in this frequency range. Higher-frequency operation is possible, and CW power levels of about 50 mW at a frequency of 200 GHz should be attainable, but at the present time there are few applications at frequencies above 100 GHz and little development is being carried out in this area. At frequencies in the 50–100 GHz range, IMPATT oscillators are being investigated for use as the primary sources in waveguide trunk communication systems and as pump sources for parametric amplifiers.

Germanium IMPATTs were investigated briefly between 1966 and 1968. At that time the early results on Ge showed promise of higher efficiency and lower noise performance compared with Si devices [26]. The improved efficiency and noise performance of Ge was attributed to the fact that the ionization rates for holes and electrons differed by only a factor of two, compared with thirty for Si. These advantages are, however, considerably outweighed by the low power-handling capability of Ge, limited by the small energy-band gap of the material. The experimental work in Ge was carried out, almost entirely, at C–X bands using diffused-junction n^{+}–p devices. The highest efficiencies reported were of the order of 12 per cent with maximum CW output powers of about 600 mW [14].

The most significant recent developments in IMPATT oscillators at low frequencies (< 20 GHz) have been the achievement of high efficiencies, high powers, and low noise from GaAs devices. The equal ionization rates and high electron mobility of this material have led to the highest efficiency CW solid-state oscillators in this frequency range. Coupled with the large energy-gap and consequent high power capability, this has led to the highest-power CW solid-state sources. The first high-power GaAs devices were reported in 1968 by Kim and Armstrong [27]. Since that time, the power and efficiency of GaAs devices have been steadily increased as the material and device technology has improved. At the time of writing the highest CW efficiency [28] obtained from GaAs IMPATTs is 19 per cent. Output powers of 3 watts at 6 GHz and 2 watts at 9 GHz have been reported for single-chip GaAs IMPATTs mounted on copper heat sinks. Increased powers up to 8 watts have been obtained using ring diodes on diamond heat sinks [29], and it seems probably that CW output powers of 10 watts at C and X band will soon be available. The GaAs device has been developed largely using the Schottky-barrier diode. As a result of the large surface-state density in GaAs, high-quality Schottky barriers can be made more easily than on Si. The high surface-state density tends to have a stabilizing effect on the properties of the semiconductor surface and helps to keep leakage currents small, enabling sharp and stable reverse breakdown characteristics to be obtained. Because the surface properties are so greatly determined by the surface-state density, the electrical properties of the metal–GaAs barrier, such as the barrier height, are not greatly affected by the particular metal used. The choice of metal is based largely on its chemical and metallurgical properties and various metals, including Pt, Mo, Ni, Ti, Cr, Pd, and nichrome, have been used successfully.

References

1. Burger, R. M. and Donovan, R. P. *Fundamentals of silicon integrated device technology. Vol. 1.* Prentice Hall Inc. New York (1967).
2. Trumbore, F. A. Solid solubilities of impurity elements in germanium and silicon. *Bell Syst. Tech. J.* **39**, 205 (1960).
3. Sze, S. M., Lepselter, M. P. and MacDonald, R. W. Metal–semiconductor IMPATT Diode. *Solid-St. Electron.* **II** (1968).
4. Zettler, R. A. and Cowley, A. M. Batch fabrication of integral heat sink IMPATT diodes. *Electron. Lett.* **5** no. 26 (1969).
5. Josenhans, J. G. Diamond as an insulating heat sink for a series combination of IMPATT diodes. *Proc. IEEE.* **56**, 762 (1968).

6. Gibbons, G. and Misawa, T. Temperature and current distribution in an avalanching p–n junction. *Solid-St. Electron.* **II**, 1007 (1968).
7. Haitz, R. H., Stover, H. L. and Tolar, N. J. A Method of heat flow resistance measurements in avalanche diodes. *IEEE Trans. Electron Devices.* **ED–16**, 438 (1969).
8. Lepselter, M. P. Beam–lead technology. *Bell Syst. tech. J.* **45**, 233 (1966).
9. Decker, D. R., Dunn, C. N. and Frost, H. B. The effect of injecting contacts on avalanche diode performance. *IEEE Trans. Electron Devices.* **ED–18**, 141 (1971).
10. Edwards, R., Cicolella, D. F., Misawa, T., Iglesias, D. E. and Decker, V. Millimetre wave silicon IMPATT diodes. International Electron Device Meeting. Oct. 1969.
11. Scharfetter, D. L., Evans, W. J. and Johnson, R. L. Double–drift–region ($p^+pn\,n^+$) avalanche diode oscillators. *Proc. IEEE.* 58, 1131 (1970).
12. Seidel, T. E., Davis, R. E. and Iglesias, D. E. Double drift-region ion-implanted millimetre wave IMPATT diodes. *Proc. IEEE.* **59**, 1222 (1971).
13. Scharfetter, D. L. and Gummel, H. K. Large signal analysis of silicon Read diode oscillator. *IEEE Trans. Electron Devices.* **ED–16,** 64 (1969).
14. Iglesias, D. E. Circuit for testing high-efficiency IMPATT diodes. *Proc. IEEE.* **55**, 2065 (1967).
15. Swan, C. B. IMPATT oscillator performance improvement with second harmonic tuning. *Proc. IEEE.* **56**, 1616 (1968).
16. Gummel, H. K. and Blue, J. L. A small signal theory of avalanche noise in IMPATT diodes. *IEEE Trans. Electron Devices.* **ED–14**, 569 (1967).
17. Misawa, T. CW Millimetre-wave IMPATT diodes with nearly abrupt junctions. *Proc. IEEE.* **56**, 234 (1968).
18. Kurakawa, K. and Magalhaes, F. M. A single tuned oscillator for IMPATT characterization. *Proc. IEEE.* **58**, 831 (1970).
19. Kenyon, N. Private communication.
20. Rucker. C. T. A multiple-diode high-average-power avalanche-diode-oscillator. *IEEE Trans. Microwave Theory & Tech.* **17**, 1156 (1969).
21. Kurakawa, K. and Magalhaes, F. M. An X-band 10 W multiple IMPATT oscillator. *Proc. IEEE.* **59**, 102 (1971).
22. Misawa, T. Microwave Si avalanche diode with nearly-abrupt type junction. *IEEE Trans. Electron Devices.* **ED–14**, 580 (1967).
23. Swan, C. B. Improved performance of silicon avalanche oscillators mounted on diamond heat sinks. *Proc. IEEE.* **55**, 1617 (1967).
24. Udelson, B. J. and Ward, A. L. Computer comparison of $n^+p\,p^+$ and $p^+n\,n^+$ junction silicon diodes for IMPATT oscillators. *Electron. Lett.* 7, 723 (1971).
25. Ying, R. S., X Band $n^+n\,p^+$ IMPATT diode. *Electron. Lett.* **8**, 297 (1972).
26. Rulison, R. L., Gibbons, G. and Josenhans, J. G. Improved performance of IMPATT diodes fabricated from Ge. *Proc. IEEE.* **55**, 223 (1967).
27. Kim, C. and Armstrong, L. D. High-power and high-efficiency GaAs avalanche diodes. *Appl. Phys. Lett.* **14**, 270 (1969).
28. Huang, H. C. *et al.* Paper presented at the IEEE International Electron Device Meeting, Washington, October 1971.
29. Irvin, J. C. Private communication.
30. Edwards, R. Private communication.
31. G. Gibbons, J. J. Purcell, P. R. Wickens and S. Gokgor, 50 GHz GaAs IMPATT Oscillator, *Electronics Lett.* 8. 513, 1972.

7 TRAPATT oscillators

7.1 Mode of operation

Although the first IMPATT oscillators were operated under pulsed-bias conditions in 1964, almost no further attention was given to the pulsed mode of operation until 1967, when the high-efficiency, high-power TRAPATT mode was discovered. In the IMPATT mode, the temperature rise at the junction, under CW bias, has little effect on the efficiency of the oscillator, and so no significant improvement in efficiency is expected for pulsed operation. Furthermore, although higher peak power is attainable under pulsed bias at low frequencies, the maximum power is quite severely restricted by the transit-time limitations. These limitations can be avoided to some extent, as shown by Gilden and Moroney [1], if an IMPATT diode is made to oscillate close to its avalanche resonance frequency. Under these conditions, the effective capacitance of the diode is considerably less than the capacitance of the depletion layer, and larger area devices can then be used.

However, with the discovery of the TRAPATT mode in 1967 by Prager, Chang, and Weisbrod [2], the prospects for constructing high-power, high-efficiency pulsed oscillators improved considerably. For about one year following the initial experimental work, the mechanism of the oscillation was not at all understood and was referred to as the 'anomalous mode'. It was, nonetheless, easily identified experimentally by several characteristic features, including:

(a) Very high d.c.-to-r.f. conversion efficiencies (up to 60 per cent),
(b) Oscillation at frequencies well below the transit-time frequency of the diode,
(c) A sudden decrease in the voltage across the diode and a consequent increase in bias current at the onset of oscillation,
(d) High threshold power for oscillation.

Thus, although experiments in several laboratories were able to produce avalanche-diode oscillators possessing these characteristics, little light was shed on the mechanisms involved in the oscillation until, in 1968, Johnson, Scharfetter, and Bartelink [3] succeeded in simulating the anomalous mode using the avalanche-diode large-signal computer programme described in §4.8. This computer simulation experiment was carried out by precisely modelling the diode structure and impressing the voltage waveform measured on an 'anomalously' oscillating device. From this information, the computer was programmed to calculate the field and carrier distributions within the active region as a function of time. The results revealed a new transient mode of breakdown of a p–n junction when a very rapidly increasing voltage is applied. An avalanche zone propagates through the depletion layer and produces a dense plasma of holes and electrons which switches the diode into a low-voltage, high-current state. The space-charge density in the plasma is considerably larger than the ionized impurity density and causes the field to collapse to a low level approaching zero. As a result, the electron and hole drift velocities drop well below the saturated values and the carriers are trapped. This is referred to as the trapped-plasma state. As the carriers drift slowly out of the active region, the electric field within the diode recovers and the voltage gradually increases. The voltage is fully recovered just as the last remaining holes and electrons are swept from the active region. The current then drops back to a low level. Thereafter, the process repeats so that the diode cycles between a high-voltage, low-current state and a low-voltage, high-current state. The current and voltage waveforms produced are extremely favourable to the production of high conversion efficiencies. The period of oscillation is much longer than in the IMPATT mode, since the electrons and holes spend a considerable part of the cycle moving at velocities well below the saturated drift velocity, and consequently the oscillation frequency is well below the Read transit frequency of the diode. Since the voltage drops to a low value approaching zero during a large portion of each cycle, the mean voltage on the diode is considerably less than the breakdown voltage. This accounts for the sudden decrease in bias voltage observed at the onset of the high-efficiency oscillations. The accompanying increase in bias current, characteristic of this mode, reflects the periodic high-current state which occurs in each cycle as the trapped plasma is formed.

7.2 The avalanche shock front and trapped-plasma formation

The new transient mode of breakdown [4] in the p–n junction, which produces the trapped-plasma state and is basic to the TRAPATT mode, is induced when a voltage greater than the static breakdown voltage of the diode is applied rapidly across the junction. An avalanche zone transits, with high velocity, across the device from the p–n junction to the base contact, and the trapped plasma is formed in the wake of this shock front. The formation of the avalanche shock front is illustrated in Fig. 7.1, which shows schematically the electric field and carrier densities in the active region of the diode at several points in time following application of the voltage. At time t_1, the applied voltage is just below the breakdown value, and the depletion layer has penetrated almost to the heavily doped substrate. As the voltage increases, the field curve moves upwards and a large displacement current flows in the active zone:

$$J = \epsilon \frac{\partial E}{\partial t}. \tag{7.1}$$

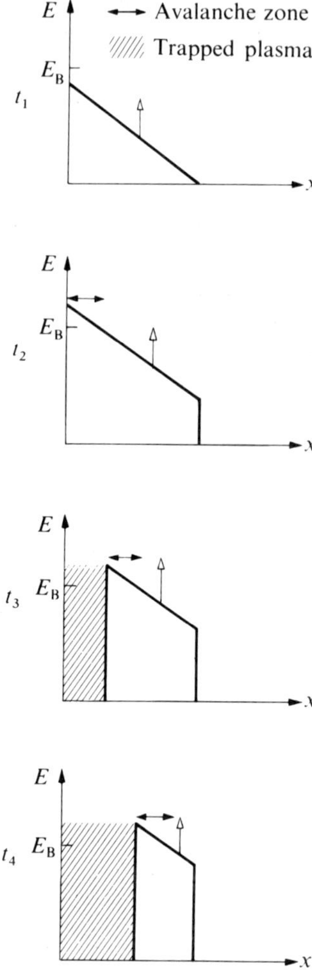

Fig. 7.1 The avalanche transient and trapped plasma formation.

Current continuity is maintained in the neutral regions by the flow of carriers swept from the active layer. At a later time t_2, the field at the junction has increased above the static breakdown value, but the diode has not yet responded because of the inherent time delay in the build-up of charge by avalanche multiplication. The field decreases abruptly to zero at the right-hand side of the active region as the depletion layer sweeps into the substrate. At time t_3 the avalanche has turned on, and a very dense hole–electron plasma is generated by the large excess field. The holes and electrons begin to drift apart, but the space charge of the mobile carriers reduces the electric field almost to zero and the carrier velocity drops to a low value determined by the low-field mobility.

After the field has collapsed, the displacement current in the plasma is zero, the current being carried entirely by the holes and electrons. In the carrier-free region, in front of the plasma, current continuity is maintained by a large displacement current which gives rise to a rapidly increasing field. Thus, at a later time t_4, the avalanche zone has advanced further to the right and the inherent delay in the charge multiplication allows the field, once again, to rise well above the static breakdown value and produce a dense hole–electron plasma. In this way the avalanche zone advances rapidly to the right, filling the entire active region with the trapped plasma. The gradient of the electric field profile across the active zone is proportional to the doping level N:

$$\frac{\partial E}{\partial X} = \frac{q}{\epsilon} N. \tag{7.2}$$

The velocity of the shock front U is given by the velocity of the intersection point of the field curve with the constant field value E_B

$$U = \frac{\partial E}{\partial t} \bigg/ \frac{\partial E}{\partial x}, \tag{7.3}$$

which, from (7.1) and (7.2), is

$$U = J/qN. \tag{7.4}$$

A typical value for the shock front velocity is 3×10^7 cm s^{-1} (about three times the saturated carrier drift velocity) and a typical value for the doping level is 4×10^{15} cm^{-3} for 1–3 GHz TRAPATTs, giving a current density of 2×10^4 A cm^{-2}. This current is about two orders of magnitude larger than the current density required in the IMPATT mode at the same frequency.

7.3 Plasma extraction

The avalanche transient or plasma formation period is followed by a longer period during which the charge is extracted from the diode and the voltage recovers. The current flowing in the device during this part of the cycle need not necessarily be equal to the current flowing during the avalanche transient period. However, for simplicity, we shall consider the special case where these currents are equal, as illustrated by the square wave in Fig. 7.2. This is an idealized current waveform which, as pointed out by Carroll and Credé [5], is only one of many that may arise in the TRAPATT mode. However it is a particularly useful model for analysing the physics of the TRAPATT diode. We shall consider an n^+–p–p^+ structure. Recovery starts at the junction side of the p-layer by the extraction of electrons, and at a slightly later time (w/U), hole extraction begins at the p–p^+ interface. At the start of the extraction period, the hole–electron densities in the active region are large and the field

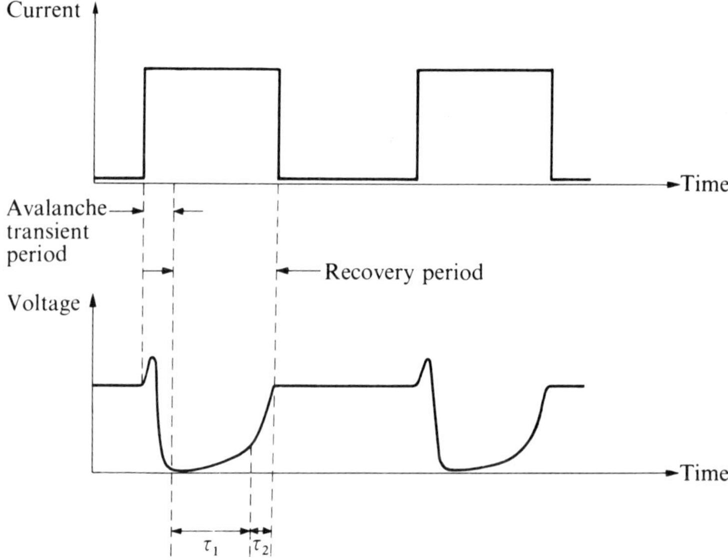

Fig. 7.2 Current and voltage waveforms for the TRAPATT oscillator.

is low everywhere, as shown in Fig. 7.3. Under the influence of the residual electric field, electrons drift to the left and holes to the right, and as the charge is removed from the active region the electric field gradually recovers. Sometime after the extraction has started, the active region is made up of three zones, as shown in Fig. 7.4. In the centre zone, the field remains low and the carrier drift velocities are determined by the low-field mobilities. In the end zones, the residual carrier densities are small and the drift velocity is given by the high-field saturated value. As the extraction continues, the boundaries between the plasma and residual zones move towards the centre of the active

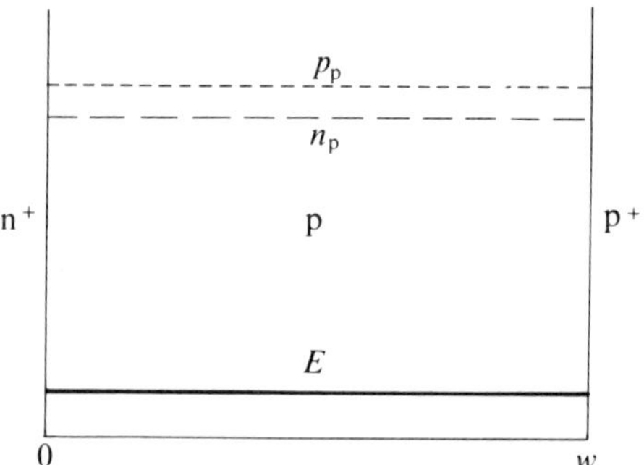

Fig. 7.3 Carrier concentrations and electric field profile in the trapped plasma at the start of the recovery period. (after Clorfeine *et al.* [6]).

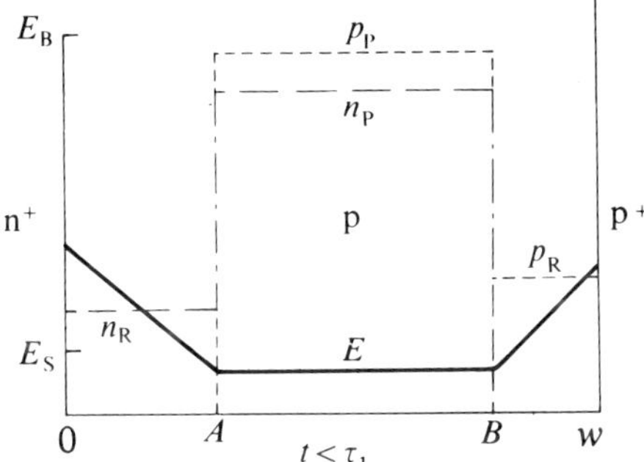

Fig. 7.4 Carrier concentrations and electric field profile during the first stage ($t < \tau_1$) of the recovery period. (After Clorfeine *et al.* [6]).

region. The velocities of these boundaries are equal to the low-field drift velocities of the holes and electrons in the plasma. As the plasma zone in the centre shrinks, the voltage on the diode increases and can be calculated by integrating Poisson's equation. The voltage waveform across the diode terminals is illustrated in Fig. 7.2.

When the field throughout the active region is restored to the value required to produce velocity saturation, as shown in Fig. 7.5, the total voltage on the diode is still well below the breakdown value. The residual carriers are then extracted at the saturated drift rate and the voltage recovers quite rapidly to about the breakdown value, as illustrated in Fig. 7.2.

Before going on to a more detailed consideration of the avalanche transient and extraction processes illustrated in Fig. 7.3 to 7.5, we must understand the triggering mechanisms necessary to initiate and sustain the TRAPATT oscillation. These are discussed in the next section.

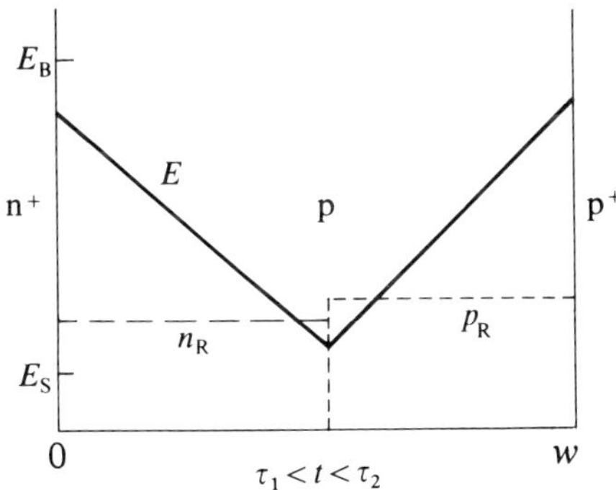

Fig. 7.5 Carrier concentrations and electric field profile during the second stage ($\tau_1 < t < \tau_2$) of the recovery period (after Clorfeine *et al.* [6]).

7.4 Triggering mechanisms

In Chapter 4 we showed that the small-signal negative resistance of the IMPATT diode leads to a natural build-up of oscillation when the diode is embedded in a resonant circuit in which the positive load resistance is smaller than the magnitude of the negative resistance of the device. The TRAPATT oscillator is quite different in that it possesses no small-signal negative resistance at the TRAPATT frequency. It is entirely a large-signal phenomenon which requires a periodic triggering pulse in each cycle to drive the diode into the trapped-plasma state. For a free-running oscillator, the trigger pulse is generated by the circuit.

The TRAPATT oscillator circuit possesses several unique features. In addition to providing a periodic voltage impulse to sustain the oscillation, the circuit must be able to support many higher harmonics of the fundamental frequency in order to shape the current–voltage waveforms for high efficiency. In practice these properties are quite easily realized in a multiple slug coaxial line similar to the IMPATT test circuit described in §6.3. The diode is located at the end of the coaxial line and bias is inserted through a standard bias tee. This configuration is essentially identical to the set-up described earlier for the IMPATT oscillator. The difference in practice between the IMPATT and TRAPATT circuits lies in the number and positions of the movable coaxial tuning slugs. In the TRAPATT circuit there are usually three or more such tuning elements. The slug nearest the diode is positioned at a distance $\lambda/2$ from the device and is responsible for providing the periodic triggering voltage at the device terminals. The remaining slugs are positioned to optimize the transfer of power at the TRAPATT frequency to the load.

The operation of the oscillator is as follows: as the bias is increased on the diode, IMPATT oscillations are initiated. The IMPATT oscillation is contained within the local cavity immediately surrounding the diode and builds up to a large amplitude. It is this oscillation which provides the voltage impulse, at the diode terminals, necessary to trigger the first trapped-plasma state. The diode voltage drops as the shock front transits and a negative voltage pulse is propagated down the transmission line, where it is reflected at the first low-impedance tuning section. On reflection, the polarity of the voltage pulse is inverted and so a positive voltage is reflected back towards the diode and arrives there one full period after the first trapped-plasma state was triggered. The reflected voltage pulse provides the over-voltage necessary to trigger a new avalanche shock front, which causes the diode voltage to collapse, and the cycle repeats.

We have, then, two triggering mechanisms involved, namely the initial voltage impulse provided by the IMPATT oscillation, which starts the TRAPATT mode, and the periodic voltage impulse provided by the circuit, which sustains the free-running oscillator.

It is also possible to operate the device as an amplifier or triggered oscillator by injecting an a.c. triggering signal either at the TRAPATT frequency or at a higher frequency.

7.5 Trapatt device theory

In this section we shall show, by considering the plasma and recovery regions in more detail, that a punch-through diode is required for efficient TRAPATT operation. The upper-frequency limits and a power–frequency relation for TRAPATTs are also presented. We shall consider a simplified model in which ionization rates, mobilities, and saturated drift velocities of holes and electrons are taken to be equal, and a square wave current as shown in Fig. 7.2 will be assumed. We must bear in mind that a square wave contains an infinite number of harmonic frequencies and is not realizable in a practical circuit. However this approximation simplifies the analysis and does not affect the essential features of the device physics. For a more detailed and general discussion of the theory of the TRAPATT oscillator, the reader is referred to the works of Clorefeine [6], De Loach and Scharfetter [7], and Cottam [8].

Sometime after the recovery period has started, the field and carrier distributions are as shown in Fig. 7.4. There are no holes in the recovery region to the left of point A and no electrons in the recovery region to the right of point B. Since charge extraction starts first the junction side of the diode ($x = 0$), the field is highest in this region.

Plasma Region

Let p_p and n_p be the hole and electron densities in the plasma, and let us assume that the particle velocity in the plasma is given by the low-field value

$$v_p = \mu E_p,$$

where E_p is the plasma field and μ is the low-field mobility, which we shall assume to be the same for holes and electrons. In the plasma region the field is flat so that, from Gauss's law, the net charge density is zero:

$$|p_p| = |N_A + n_p|. \tag{7.5}$$

The current in the plasma region is entirely particle current and is equal to the current I through the device:

$$I = qv_p(p_p + n_p). \tag{7.6}$$

The velocity of the avalanche shock front U is related to the current and background doping density as given in (7.4). Equations (7.5), (7.6), and (7.4) determine the electron and hole concentrations in the plasma, namely

$$p_p = \frac{N_A}{2}\left(\frac{U}{v_p} + 1\right), \tag{7.7}$$

$$n_p = \frac{N_A}{2}\left(\frac{U}{v_p} - 1\right). \tag{7.8}$$

The transit time for the shock front is

$$\tau_A = \frac{w}{U}. \tag{7.9}$$

Recovery Region

Let n_R and p_R be the electron and hole concentrations in the recovery region, and we shall assume that these carriers are moving with the saturated drift velocity. We refer back to Fig. 7.4 and 7.5 which illustrate the two stages in the recovery period. During the first part of the period ($t < \tau_1$) the charge is extracted at the edges of the active region, and the field rises above the saturation value. The trapped-plasma zone in the centre of the active region is shrinking as the points A and B move to the right and left respectively. The velocity of points A and B are equal and given by the carrier velocity in the plasma v_p. The first stage of the recovery period lasts until A and B meet at a time τ_1 after the start of charge extraction. In a reference frame moving with the velocity of point A, we can equate the electron flow in the recovery region with the electron flow in the plasma:

$$n_R(v_p + v_s) = 2n_p v_p. \tag{7.10}$$

The points A and B meet after a time

$$\tau_1 = w/2\,v_p; \tag{7.11}$$

thereafter the charge is extracted at the saturated drift velocity, and so the recovery is complete after a further time

$$\tau_2 = w/2v_s \tag{7.12}$$

has elapsed. Thus the total extraction time τ_E is given by

$$\tau_E = \frac{w}{2}\left[\frac{1}{v_p} + \frac{1}{v_s}\right]. \tag{7.13}$$

During the recovery period, the field in the active region is rising, and for efficient TRAPATT operation it is required that the maximum field in the device should not rise above the static breakdown field E_B at the end of the extraction period. The field is given by

$$\frac{dE}{dt} = v_p \frac{dE}{dy}, \tag{7.14}$$

since point A is moving with velocity v_p. Poisson's equation in the recovery region is

$$\frac{dE}{dy} = -\frac{q}{\epsilon}(n_R + N_A), \tag{7.15}$$

and so the maximum field in the device ($x = 0$) at the end of the extraction period is given by

$$E = E_{\mathrm{p}} + \frac{q v_{\mathrm{p}} \tau_{\mathrm{E}}}{\epsilon} [n_{\mathrm{R}} + N_{\mathrm{A}}]. \tag{7.16}$$

This must be less than or equal to the breakdown field E_{B}. Substituting from equations (7.10), (7.8), and (7.13), we get

$$E_{\mathrm{B}} - E_{\mathrm{p}} \geqslant \frac{qw}{2\epsilon}\left[\frac{1}{v_{\mathrm{p}}} + \frac{1}{v_{\mathrm{s}}}\right] v_{\mathrm{p}} \left[\frac{v_{\mathrm{p}} N_{\mathrm{A}}}{v_{\mathrm{p}} + v_{\mathrm{s}}}\left(\frac{U}{v_{\mathrm{p}}} - 1\right) + N_{\mathrm{A}}\right] \tag{7.17}$$

which simplifies, using $U \ll v_{\mathrm{p}}$, $v_{\mathrm{s}} \gg v_{\mathrm{p}}$, $E_{\mathrm{B}} \gg E_{\mathrm{p}}$, to give

$$\frac{2\epsilon E_{\mathrm{B}}}{qWN_{\mathrm{A}}} \geqslant \frac{U}{v_{\mathrm{s}}} + 1. \tag{7.18}$$

Now $\epsilon E_{\mathrm{B}}/qN_{\mathrm{A}}$ is equal to the depletion layer width w_{B} of the non-punch-through abrupt-junction diode with doping level N_{A}. Therefore,

$$\frac{2w_{\mathrm{B}}}{w} \geqslant \frac{U}{v_{\mathrm{s}}} + 1. \tag{7.19}$$

Now, for the avalanche shock front to propagate, U must be greater than v_{S}, and from (7.19) w must be less than w_{B}. Thus, to avoid premature avalanching at the end of the recovery period, a punch-through diode is required. This is an important conclusion of the theory and is borne out by experimental results. Equation (7.19) may be written

$$2F \geqslant \frac{U}{v_{\mathrm{s}}} + 1, \tag{7.20}$$

where F is the punch-through factor defined in §3.1. In a typical TRAPATT diode $F \approx 2$ and $U \approx 3V_{\mathrm{S}}$.

We have assumed, in the above analysis, that the low-field and saturated-drift velocities of holes and electrons are equal. This assumption simplifies the analysis but does not significantly affect the device physics. When the velocities are not equal, the extraction time is determined by the carrier with the smaller velocity. We have also assumed here that the current flowing during the avalanche transient period is equal to the current flowing during the extraction period. This is a special case which again simplifies the analysis. A more general current waveform is one in which the current during the avalanche transient period is larger than the current flowing during the extraction period. This type of waveform arises when the capacitance of the diode and its package are included. The capacitance charges during the low-current high-voltage part of the cycle, and discharges during the avalanche transient period, producing a larger current. Since the charge is extracted at a lower current, the recovery time is longer in this case, and the frequency of the oscillator is lower.

Frequency of operation

For efficient operation of the TRAPATT oscillator, the duration of the high-current, low-voltage state should be equal to the duration of the low-current, high-voltage state. Therefore,

$$\tau_A + \tau_E = \frac{1}{2f}, \tag{7.21}$$

where f is the frequency of oscillation which, as we have already seen, is circuit-controlled by the length of line between the diode and the first low-impedance tuning slug. The left-hand side of (7.21) is determined by the diode structure and bias current. Given a diode structure in which N_A and w are specified, the shock front velocity U is determined from (7.18), provided the breakdown field and dielectric constant of the material and the saturated carrier velocity are known. The current I is then found from (7.4). Using (7.9) and (7.13), (7.21) becomes

$$w\left[\frac{1}{U} + \frac{1}{2v_p} + \frac{1}{2v_s}\right] = \frac{1}{2f}; \tag{7.22}$$

and to find the optimum frequency for any structure, we must know the plasma velocity. The field in the plasma region E_p, and therefore the plasma velocity $v_p = \mu E_p$, can be calculated using Poisson's equation and the continuity equation, provided the current is known. The derivation is not presented here.

An upper limit to the frequency of operation for any structure is set by

$$f_{max} = \frac{1}{\tau_A + \tau_E}, \tag{7.23}$$

which is just twice the optimum frequency. Cottam [8, 9] has calculated the optimum and maximum frequencies for both Si and GaAs TRAPATTs. The results for Si p^+–n–n^+ devices are shown in Figs. 7.6 and 7.7.

Power–frequency relation

In a more general analysis, Scharfetter [10] has calculated the power–frequency characteristics for TRAPATT oscillators in both Ge and Si, without making the assumption of equal current flow during the avalanche transient and extraction periods. The d.c.-to-a.c. conversion efficiency is calculated by Fourier analysing the current and voltage waveforms, and the total output power is calculated by normalizing the diode area, in such a way that the negative resistance of the device is 10 ohms in each case. The calculations were performed for a specific case, in which the TRAPATT frequency was taken to be one-third of the nominal IMPATT frequency of the device (transit angle π). The results for Si devices are shown in Figs. 7.8–7.10. In Figs. 7.8 and 7.9, the depletion-layer width w and the doping density N_A are plotted versus frequency. From these

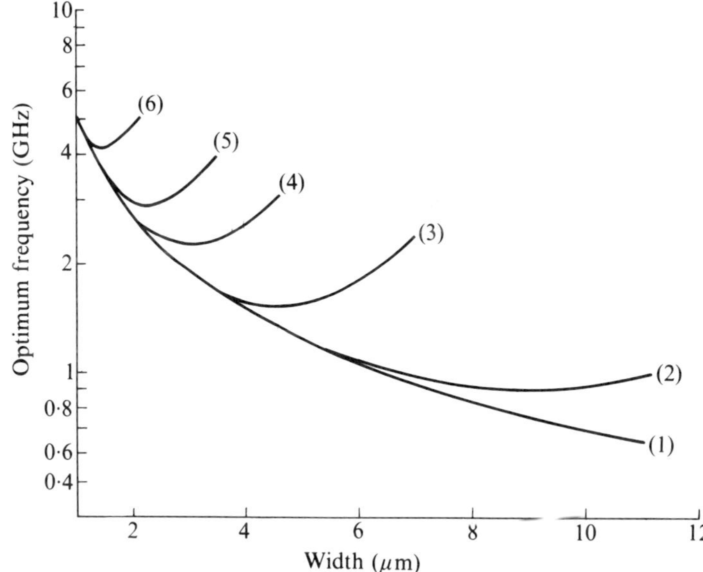

Fig. 7.6 Optimum frequency of Si p^+–n–n^+ TRAPATT diodes versus depletion layer width for different doping densities (after Cottam [9]).
(1) 6×10^{14}cm^{-3}, (2) $1{\cdot}5 \times 10^{15}$cm^{-3}, (3) 2×10^{15}cm^{-3}, (4) 3×10^{15}cm^{-3}, (5) 4×10^{15}cm^{-3}, (6) 6×10^{15}cm^{-3}

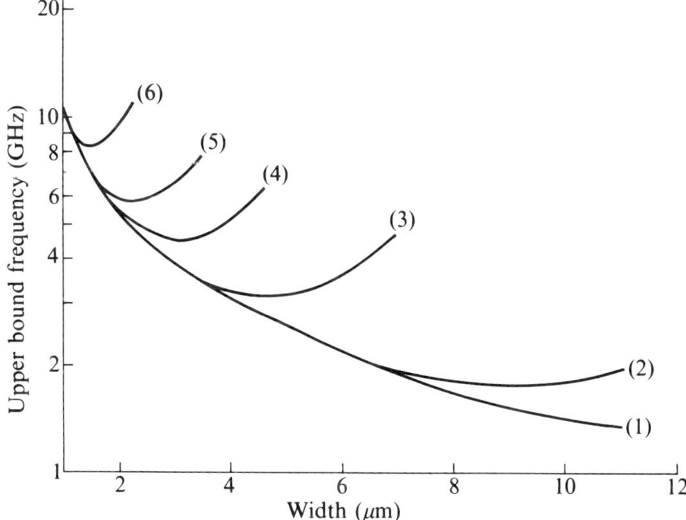

Fig. 7.7 Upper bound frequency of Si p^+–n–n^+ TRAPATT diodes versus depletion-layer width for different doping densities (after Cottam [9]).
(1) 6×10^{14}cm^{-3}, (2) $1{\cdot}5 \times 10^{15}$cm^{-3}, (3) 2×10^{15}cm^{-3}, (4) 3×10^{15} p^+–cm^{-3}, (5) 4×10^{15}cm^{-3}, (6) 6×10^{15}cm^{-3}.

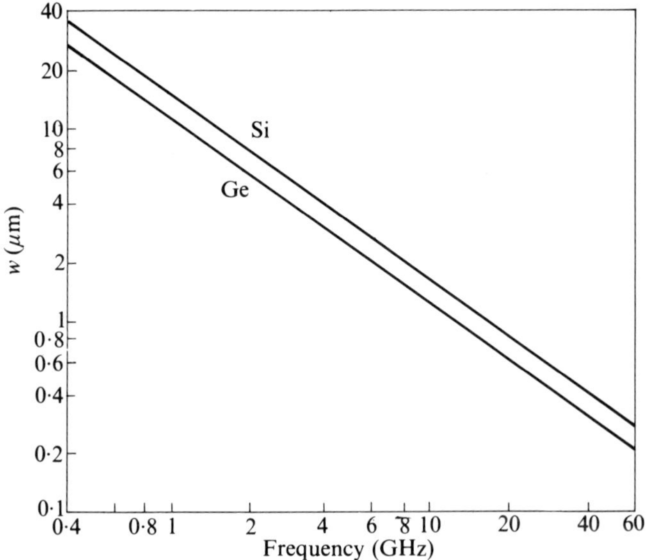

Fig. 7.8 Depletion-layer width w versus frequency for Ge and Si TRAPATT divides (after Scharfetter [10]).

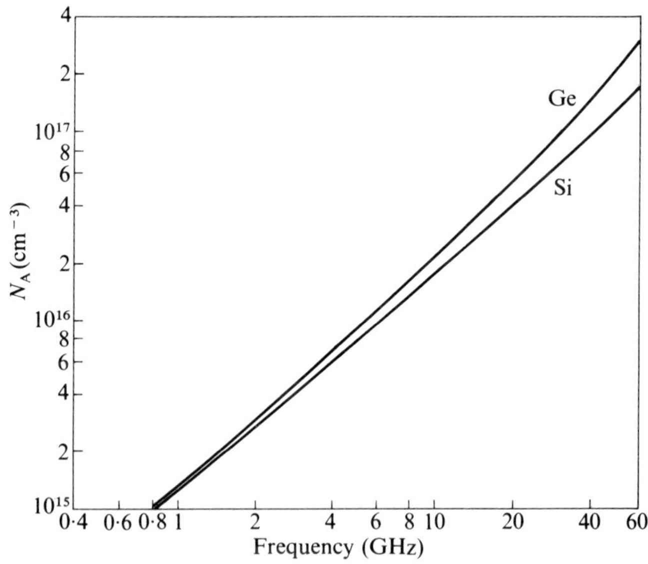

Fig. 7.9 Doping density N_A versus frequency for Ge and Si TRAPATT diodes (after Scharfetter [10]).

results, along with the depletion-layer width versus doping density in the non-punch-through diode (Fig. 3.8), the punch-through factor F at any frequency can be calculated. The microwave output power versus frequency is shown in Fig. 7.10 for both Si and Ge devices. The lower power capability of Ge reflects the small band-gap and low breakdown field of that material. The CW limit, based on realistic thermal impedance, is shown by the dotted line. At low frequencies, below about 10 GHz for Si, the thermal limitations restrict the diode size and CW output power, and at these frequencies increased power can be obtained under pulsed operation. Above 10 GHz the performance of Si TRAPATTs is determined by electronic rather than thermal limitations, and the same power level is obtained in both CW and pulsed operation.

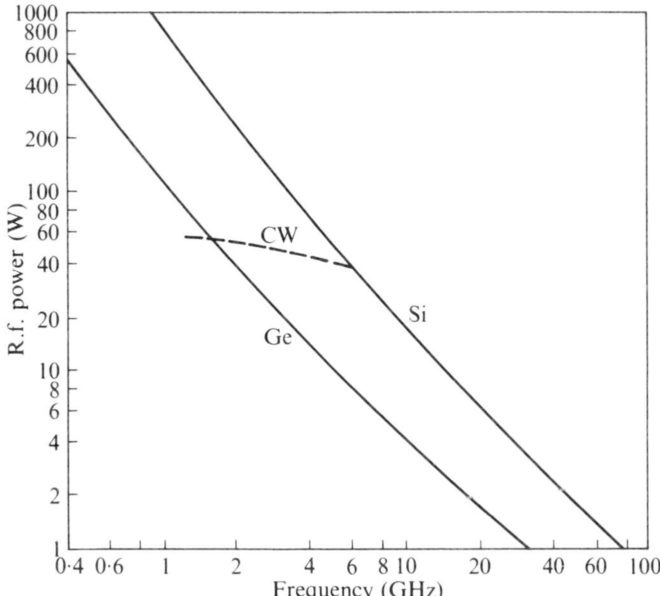

Fig. 7.10 Output power versus frequency for Ge and Si TRAPATT diodes versus frequency (after Scharfetter [10]).

7.6 Practical device design

All high-efficiency TRAPATT oscillators have been constructed using either Si or Ge diodes. The main difference between the two materials is the input power density required to sustain high-efficiency oscillation. The low breakdown field and small saturated drift velocity in Ge combine to give theshold power densities about a factor of four smaller than Si [7]. The techniques for constructing TRAPATT diodes are essentially the same as those used for IMPATT devices, and have been described in §6.1. The TRAPATT diode differs from the IMPATT in that the depletion layer punches through to the substrate before breakdown occurs.

We have seen that the punch-through factor F defined in §3.4 is a useful parameter characterizing this structure. An alternative and related parameter is the punch-through voltage V_p, which is the reverse-bias voltage required to fully deplete the epitaxial layer

$$V_p = \frac{q}{2\epsilon} N w^2 ,$$

where N is the doping density and w the width of the layer. Then V_p and F are related by

$$F = \left(\frac{V_B}{V_p}\right)^{\frac{1}{2}}, \tag{7.24}$$

where V_B is the breakdown voltage of the non-punch-through diode ($V_B = q/2\epsilon\, N w_B^2$). The significance of the punch-through voltage arises from the fact that it is this parameter, rather than F, which is most easily measured experimentally. V_p is obtained from a plot of the diode capacitance versus applied voltage under reverse bias. For voltages less than V_p, the junction capacitance decreases as $1/V^{\frac{1}{2}}$, as expected for the abrupt-junction diode in accordance with eqn (3.16). When punch-through is reached, the capacitance is just $C = \epsilon A/w$ and will remain constant despite further increase in voltage, since the depletion layer will not penetrate significantly into the n^+ substrate. Thus V_p is obtained in practice as the voltage at which the C–V characteristic becomes flat.

Another parameter useful in characterizing a TRAPATT oscillator is the ratio of the nominal IMPATT frequency of the diode (transit angle π) to the frequency of operation in the TRAPATT mode. This we will refer to as the subharmonic number. The current and voltage waveforms can be Fourier-analysed into a fundamental frequency (the TRAPATT frequency) and several harmonics, the highest being the IMPATT frequency. Waveforms favourable for efficient generation of microwave power are obtained when many harmonics are present; that is, when the subharmonic number is large.

A simple design theory for TRAPATT diodes, relating the current density to the punch-through factor, can be developed on the basis of the following criteria:

(a) Premature breakdown at the end of the recovery period must be avoided.
(b) The shock front velocity is greater than the saturated drift velocity.
(c) The width of the diode is larger than the width of the avalanche zone.
(d) The current is a square wave.

To avoid premature avalanching, the total current in the trapped-plasma state is given from (7.4) and (7.19) by

$$J \leqslant qNv_s(2F - 1). \tag{7.25}$$

Now, for a square-wave current in which the high-current, low-voltage state exists for exactly one half of the period, the d.c. current $J_{\text{d.c.}}$ is given by $J_{\text{d.c.}} = J/2$. From (7.25) we obtain

$$J_{\text{d.c.}} \leqslant \frac{qNv_s}{2}(2F - 1). \tag{7.26}$$

This result is shown graphically in Fig. 7.11, where the normalized bias current $J_{\text{d.c.}}/qNv_s$ is plotted (by the line labelled (1)) as a function of F. High-efficiency oscillations can be sustained only for points lying below this line. A lower limit to the current density is set by the condition that the shock-front velocity is greater than the saturated drift velocity. From (7.4) we have

$$J_{\text{d.c.}} \geqslant \frac{1}{2} qNv_s. \tag{7.27}$$

This limit is shown in Fig. 7.11 by the constant-current line labelled (2). An upper limit to the value of F is set by condition (c) above. When the width of the active region of the diode is less than or equal to the width of the avalanche

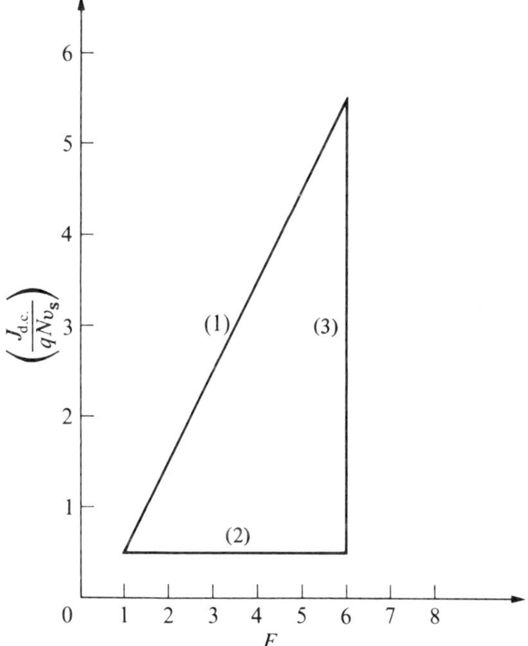

Fig. 7.11 Design triangle on $(J_{\text{d.c.}}/qN_{v_s})$ versus F plot for TRAPATT diodes.

zone, the model used in the analysis no longer applies. For a non-punch-through diode, the avalanche zone occupies about one-third of the depletion layer, suggesting $F = 3$ as an upper limit for the punch-through factor. However, larger values of F are in fact possible, since the breakdown field gradually increases and the avalanche zone shrinks as the width of the device is decreased (F increased). A somewhat more realistic upper limit for the punch-through factor is $F \simeq 6$. This is represented by the vertical line, labelled (3), in Fig. 7.11. For high-efficiency operation in the TRAPATT mode, the current density and punch-through factor must lie in the triangle in Fig. 7.11. By analysing a more general case with unequal low-field mobilities, appropriate for Si, Clorfeine [11] constructed this design triangle using $J_{\text{d.c.}}$ and the product (Nw) as coordinates. He showed that all the well-documented, high-efficiency ($\eta > 50$ per cent) experimental results for Si TRAPATTs correspond to operation within this triangle.

Although the doping density and width of the diode are important design parameters determining the efficiency of the device they do not play a particularly critical role in determining the frequency of oscillation. This is controlled by the circuit and can, in general, be varied over more than an octave, and in some cases much more by mechanically tuning the circuit. In practice the bandwidth of the oscillator is normally limited more by the circuit than by any inherent limitations of the diode itself. In one case a CW Ge device was tuned from 300 MHz to 1200 MHz [12]. On the other hand, diodes with quite different structures can operate at the same frequency in the same circuit. This was demonstrated by Gibbons and Grace [13], using three diodes, with the same value of N, but with F values of 1·5, 2·5, and 4·5, corresponding to breakdown voltages of 85 V, 68 V, and 58 V respectively. The reverse-bias capacitance–voltage plots for these three diodes are shown in Fig. 7.12. Each of these devices

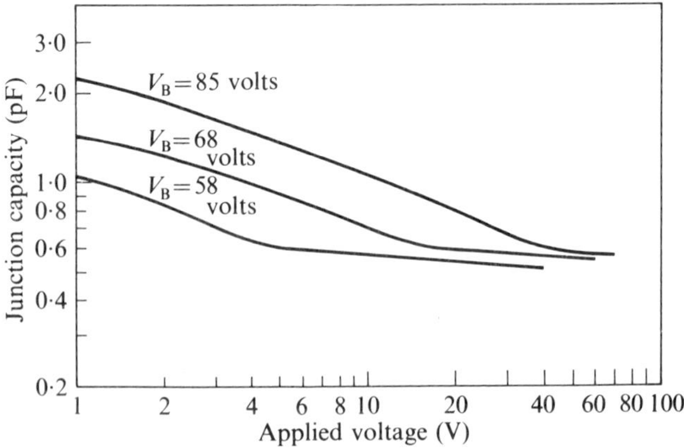

Fig. 7.12 Reverse bias capacitance versus voltage for three TRAPATT diodes measured in a coaxial circuit. (After Gibbons and Grace [13].)

oscillated in the 1 to 3 GHz frequency range, with efficiencies between 20 and 30 per cent, and one circuit configuration could be found in which all three devices operated at the same frequency without retuning. The oscillation frequency was 1·96 GHz, and was the same (to within a few MHz) for all three devices. At this frequency, the conversion efficiencies were about the same (~ 30 per cent) for the 58 V and 68 V diodes, but slightly lower (~ 25 per cent) for the 85 V. The most significant difference in the performance of these three diodes was the input power level necessary to sustain the high-efficiency oscillation. The low-voltage (large F) diode, which lies close to the left-hand side of the design triangle, required significantly less d.c. input power. This region is best suited for CW operation where low threshold power density is desired. In contrast the high-voltage diode is best suited for pulsed applications where high peak power is required.

7.7 High-frequency TRAPATTs

At frequencies above about 10 GHz the construction of TRAPATT oscillators becomes increasingly difficult. Broad-band circuits are required to support the harmonic frequencies, and for efficient operation, at least two harmonics must be present. Thus, for example, a 10 GHz oscillator circuit must provide resonances at 10 GHz, 20 GHz, and 30 GHz, and as the fundamental frequency is increased, the practical realization of such a circuit becomes more difficult. In this section we shall consider the problems associated with the diode design for high-frequency operation, following the criteria developed by Evans [14] from computer simulations of TRAPATT oscillators.

To some extent, the oscillation frequency of any diode can be increased by operating at a lower subharmonic number, but this invariably results in a lower conversion efficiency. To increase the frequency of operation, and at the same time maintain high efficiency, it is necessary to increase the nominal IMPATT frequency of the diode. This is achieved by increasing the doping density or by increasing the punch-through factor, but in either case the junction breakdown field is increased and the TRAPATT mode is then more difficult to trigger. This difficulty arises because the rate of change of the ionization rate with field tends to saturate at high fields ($d\alpha/dE \sim \alpha/E^2$ for Si and Ge) and a larger excess field is then required to produce the over-multiplication. Thus, as the frequency is increased, a larger over-voltage must be reflected back to the diode terminals to sustain the oscillation. This can be most easily achieved in a device with a large breakdown voltage (large F value). This is illustrated in Fig. 7.13, which shows field profiles for two diodes of equal width, but different values of F. The p^+–n–n^+ device has a punch-through factor of 1·4 compared with a value of 4 for the p^+–ν–n^+ diode. Since the avalanche shock front is produced when the field reaches a certain critical value, typically 1·5 E_B, and since E_B does not change significantly with doping density, the

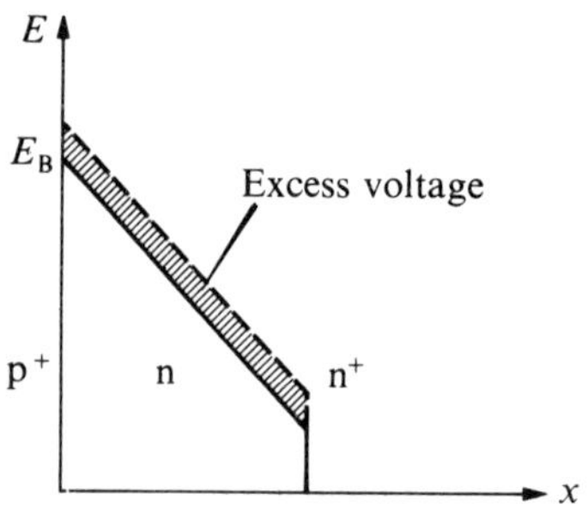

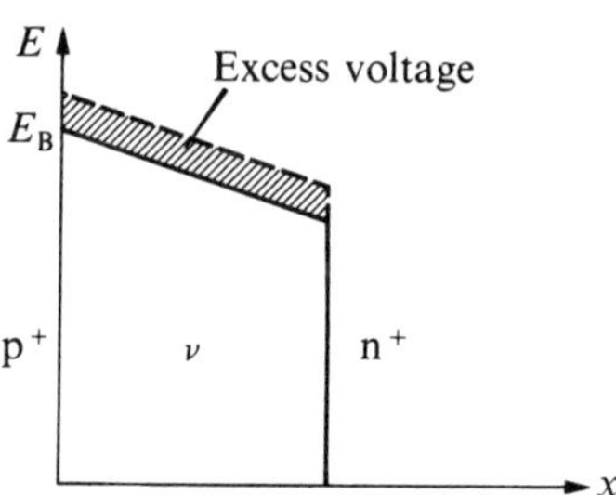

Fig. 7.13 Field Profiles for p^+–n–n^+ and p^+–ν–n^+ TRAPATT diodes.

same over-voltage is required in both cases, but the static breakdown voltage of the p^+–ν–n^+ diode is larger. A larger periodic voltage impulse can be generated by the circuit when the diode breakdown voltage is large, and the repetitive triggering is therefore more easily achieved with the p^+–ν–n^+ structure. Unfortunately, however, this structure is more difficult to start. We have seen earlier, in §4.7, that as the width of the avalanche region widens, corresponding to a large F value, the magnitude of the negative resistance at the IMPATT frequency decreases and the large-signal IMPATT oscillation necessary to produce the first trapped-plasma state cannot be developed.

The opposite situation applies in the low-voltage p^+–n–n^+ diode where the punch-through factor is small. The first trapped-plasma cycle is easily triggered by IMPATT oscillations, but the periodic over-voltage required to sustain the oscillator cannot be generated. It is apparent, therefore, that the conditions required for initial and periodic triggering cannot be obtained simultaneously in the conventional diode structures as the frequency is increased.

In an attempt to overcome these problems Evans, Seidel, and Scharfetter [15] have proposed a double-sided diode, shown in Fig. 7.14, in which the IMPATT and TRAPATT structures are fabricated separately. The TRAPATT part of the device is designed to facilitate periodic triggering, and is built back-to-back with a high-frequency IMPATT diode which provides the initial triggering voltage.

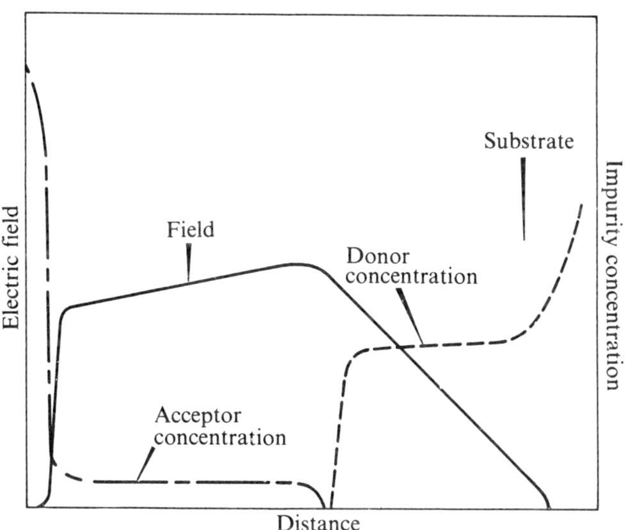

Fig. 7.14 Doping density and field profile for novel TRAPATT–IMPATT diode (after Evans *et al.* [15]).

The double-sided device has been fabricated successfully using ion-implantation techniques similar to those used for constructing the double-drift IMPATT diode described in §6.4. CW operation at 4.8 GHz has been reported [15] but higher frequencies have not yet been obtained with this structure.

7.8 TRAPATT circuits

The standard TRAPATT circuit is a 50-ohm coaxial air line with the diode mounted at one end. Movable, low-impedance tuning sections inserted between the diode and load are used to tune the frequency and optimize the output power. The diameter of the outer conductor is normally 7 mm. In a coaxial line of this type, the propagating mode is a TEM mode, provided the frequency is less than $c/\pi(b + a)$, where b is the radius of the outer conductor and a is the radius of the inner. For a 7-mm 50-ohm line, a = 1·5 mm. Thus the TEM mode alone is propagated at all frequencies up to 20 GHz. Above this frequency other modes are possible and the dimensions of the line have to be reduced if operation above 20 GHz is required.

The general characteristics of the TRAPATT circuit are those of a low-pass filter. Fig. 7.15 shows the return loss versus frequency, looking into the circuit from the diode side of the tuning slugs. These results were obtained by Evans [16] on a 500 MHz oscillator. At the TRAPATT frequency the circuit has a large resistive component and a capacitive reactance. At higher frequencies the impedance is equivalent to that looking into a short-circuited transmission line

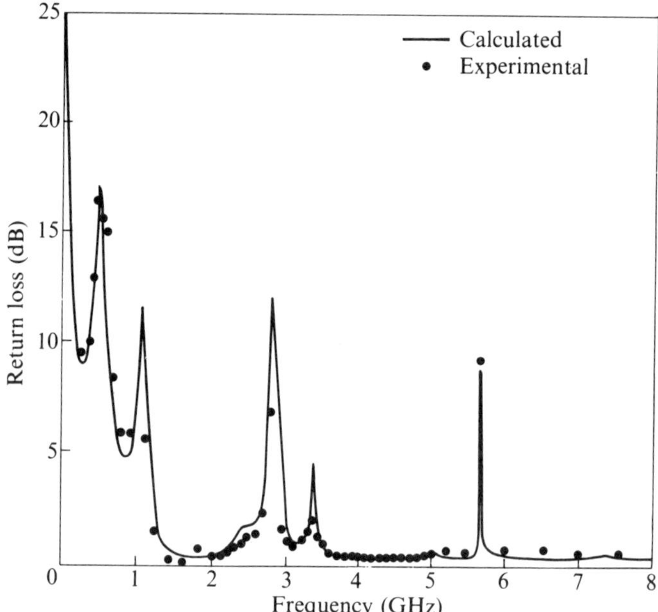

Fig. 7.15 Return loss versus frequency for a TRAPATT circuit (after Evans [16]).

one-half wavelength long at the fundamental frequency, and therefore provides a resonance at the fundamental and all its harmonics.

The fundamental frequency is determined by the length of line l between the diode and the filter section. Experimental results of frequency versus $1/l$ are shown in Fig. 7.16. These results were obtained by Evans and Scharfetter [12] on pulsed and CW oscillators at frequencies from about 300 MHz to 2·5 GHz. Notice that although the frequency is inversely proportional to the length of line, the experimental points do not lie on the line $f_0 l = 1$. The reason for this is that the round-trip delay for the pulse travelling from the diode to the filter and back is slightly less than one full period. This can be understood by referring to the voltage waveform at the diode terminals shown in Fig. 7.2. The voltage drop in the first pulse is reflected back at the diode to give the over-voltage for the second pulse. Therefore, the difference between the period of the oscillation and the total round-trip delay is equal to the time it takes for the diode voltage to collapse to zero during the avalanche transient period.

In practice TRAPATT operation has been obtained at frequencies up to about 6 or 7 GHz in the standard 7 mm coaxial line. This is about the maximum frequency to be expected from this type of circuit. Frequencies in X-band, between 8 GHz and 10 GHz, have been obtained using circuits specifically designed for higher-frequency operation. One such circuit is a miniature 50-ohm

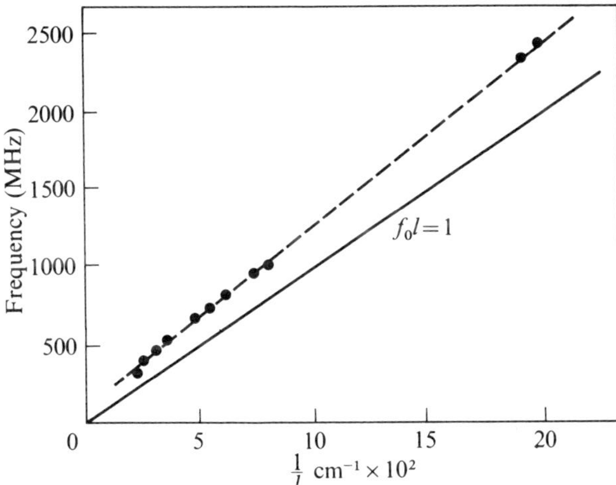

Fig. 7.16 Frequency versus $1/l$ where l is the distance between the diode and the low-impedance section. ([12].)

coaxial line with an outer conductor diameter of 4·3 mm, such that higher-order modes are cut off below 35 GHz [17]. A novel circuit for high-frequency operation is a ridged waveguide structure with movable tuning slugs inserted through the top wall of the waveguide [13] as shown in Fig. 7.17. The ridged section of K-band (18 GHz to 25 GHz) waveguide is designed to lower the cut-off frequency of the guide to 7·5 GHz. The cut-off frequency of the next higher mode is 25 GHz. The K-band section is tapered into a full height X-band waveguide. This circuit has a 3 to 1 frequency range allowing operation at the lower end of X-band with a subharmonic number of 3.

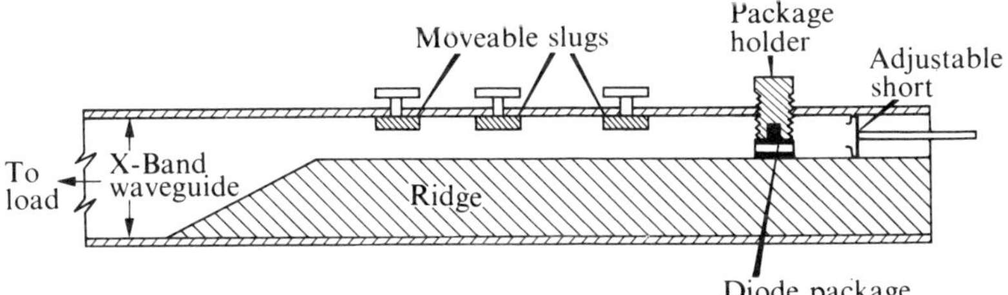

Fig. 7.17 Ridged waveguide circuit for X-band TRAPATT operation (after Gibbons and Grace [13]).

7.9 Harmonic extraction

An alternative method for increasing the frequency of the TRAPATT oscillator is by extracting the microwave power at a harmonic of the fundamental TRAPATT frequency. Adjustment of the circuit tuning conditions will allow the fundamental TRAPATT frequency to be terminated reactively while providing a resistive termination at one of the higher harmonics, usually the second or third. This can be achieved using the conventional double-slug circuit, keeping the distance between the first slug and the diode fixed, and decreasing the distance between the first and second slug. As the slug separation is decreased, the frequency for power extraction is increased. Using a diode which produced an efficiency of 35 per cent at a fundamental frequency of 1 GHz, Slaymaker *et al.* [18] were able, by this technique, to extract power at the second, third, and fourth harmonics (2,3, and 4 GHz) with efficiencies of 26, 34, and 10 per cent respectively.

A novel approach for the extraction of harmonic power has been devised by Liu [19]. The diode is embedded in a circuit which contains the usual low-impedance tuning sections between diode and load. Behind the diode is a short section of open-circuited line as shown in Fig. 7.18(a). The lengths of line between the diode and both the open circuit and the low-impedance section are chosen to be equal to $\lambda/4$ at the fundamental TRAPATT frequency. Power is then extracted at the second harmonic. Triggering of the trapped-plasma state is controlled by multiple reflections of the voltage pulse from the low-impedance section and the open circuit in turn. We must keep in mind that a voltage pulse is inverted when reflected from a short circuit but maintains the same polarity

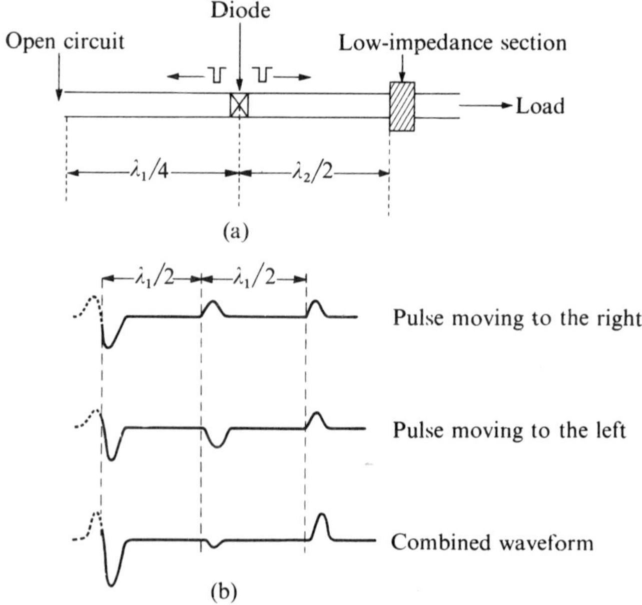

Fig. 7.18 (a) Circuit for second harmonic (λ_2) extraction and (b) voltage waveforms at the diode terminals produced by multiple reflections from the open and short circuit (after Liu [19]).

when reflected from an open circuit. Let us assume that the first trapped-plasma state has been triggered, generating a negative voltage pulse at the diode terminals. Consider first the propagation of this pulse to the right. It travels down the line to the short circuit where it is inverted and reflected back to the diode, arriving there one half period later. This positive-going voltage pulse continues to the left and is reflected from the open circuit without inversion. The pulse reflected from the open circuit arrives back at the diode after a further delay of one half period. The voltage waveform at the diode, produced by this pulse, is shown in Fig. 7.18(b). Now consider the pulse which initially travels toward the open circuit. This pulse is reflected first at the open circuit, then at the short circuit, and produces a voltage waveform at the diode terminals as shown. The combined waveform, also shown in Fig. 7.18(b), is obtained by adding these two components and produces a positive triggering pulse at the diode once each period of the fundamental TRAPATT oscillation.

7.10 Practical results

The first TRAPATT oscillators were constructed using Si diodes, and produced conversion efficiencies up to 60 per cent under pulsed-bias conditions at frequencies between 500 MHz and 1 GHz. Since these early results significant advances have been made towards increasing the frequency of operation, but little improvement has been obtained in the oscillator efficiency. This is as we expect, since 60 per cent is close to theoretical maximum efficiency in the TRAPATT mode. In addition, through improved diode design and better heat sinking, CW operation has been achieved in both Si and Ge.

The present (1972) performance data for pulsed Si TRAPATTs are shown in Fig. 7.19. Fundamental frequencies up to 10 GHz have been observed, and output powers of over 1 kW have been obtained at L-band. The highest efficiencies, 60 to 75 per cent, have been measured at the low end of the frequency range. The dotted line in Fig. 7.19 represents Scharfetter's power–frequency limit for a subharmonic number of 3, discussed earlier (see Fig. 7.10). Some improvement in power capability is to be expected, but it is apparent from these results that the state of the art is well advanced. The point lying above the theoretical limit was obtained using a series connection of five diodes. The remaining data points are for single-diode oscillators. The point lying closest to the theoretical line (19W at 8·6 GHz) does not represent a free-running oscillator. This result was obtained in an amplifier mode (external triggering) with 7 dB of gain at the saturated output power condition. All other points represent free-running oscillators. The highest-power result (1·2 kW at 1·1 GHz) was reported by Liu and Risko [20] using five diodes connected in series. This is the highest power level obtained to date from an avalanche-diode oscillator. The diode chips used in this experiment were capable, individually, of generating output powers of about 200W. In the series configuration the diodes were separately mounted in pill packages which were then stacked in a larger package. The series combination produces a high device impedance which can easily be matched in the circuit, and has the additional advantage that careful selection of diodes with equal breakdown voltage is not necessary.

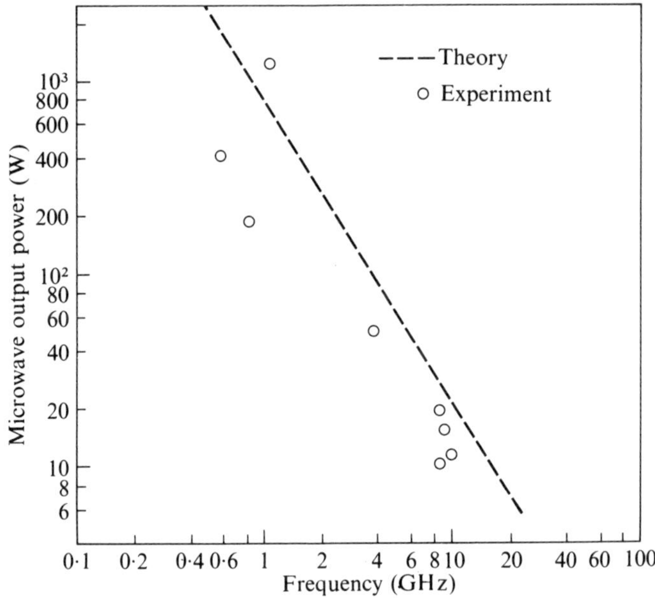

Fig. 7.19 State of the art (1972) for pulsed Si TRAPATT diodes.

The main disadvantage of the series configuration is the difficulty in removing heat from all the diodes at the same rate, and operation at a high-duty cycle is therefore precluded. The highest-duty cycle achieved by Liu and Risko at the kilowatt power level was 0·1 per cent, and the pulse length was typically 0·5 microseconds. Parallel connections of similar diodes were less satisfactory for two reasons. To ensure uniform current distribution, diodes with the same breakdown voltage must be carefully selected. The lower impedance level of the parallel array makes impedance matching difficult, and lower powers are obtained as a result.

Although the conversion efficiency is high in the TRAPATT oscillator, a substantial fraction of the d.c. input power is nevertheless dissipated as heat in the junction. The power levels required to sustain the high-efficiency oscillation are considerably above the levels used in the IMPATT mode, and as a result continuous operation in the TRAPATT mode is difficult to achieve. This is particularly true in Si, where the power density is about four times greater than that required in Ge. The current density at the oscillation threshold may be reduced by increasing the harmonic number. For this reason, the first CW operation was obtained at very low frequencies around 500 MHz [21]. Subsequent work using narrower devices to increase the harmonic number led to improved performance at higher frequencies. The state of the art (1972) for Si CW TRAPATTs is shown in Fig. 7.20, along with the power-frequency limit derived by Scharfetter.

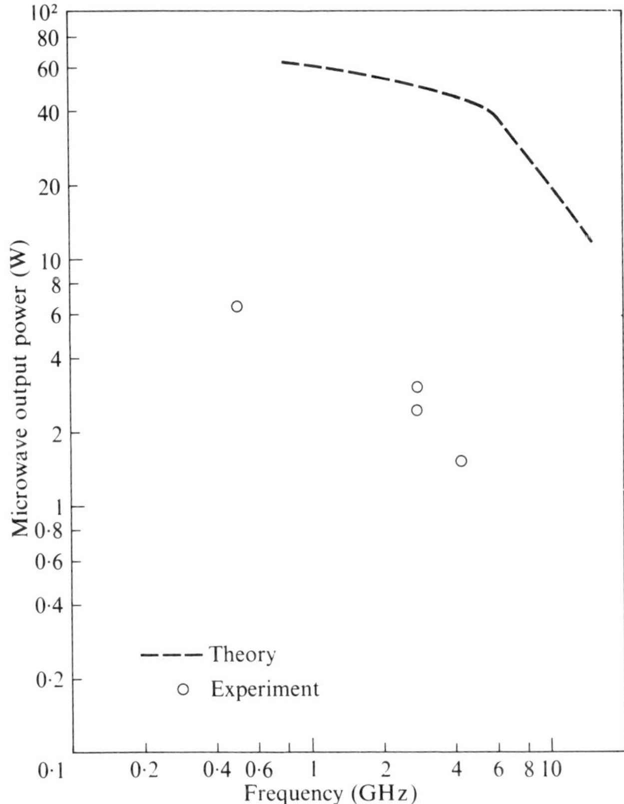

Fig. 7.20 State of the art (1972) for CW Si TRAPATT diodes.

References

1. Gilden, M. and Moroney, W. High power pulsed avalanche diode oscillators for microwave frequencies. *Proc. IEEE.* **55**, 1227 (1967).
2. Prager, H. J., Chang, K. K. N. and Weisbrod, S. High power, high efficiency silicon avalanche diodes at ultra high frequencies. *Proc. IEEE.* **55**, 586 (1967).

3. Johnson, R. L., Scharfetter, D. L. and Bartelink, D. J. High efficiency sub-transit time oscillations in germanium avalanche diodes. *Proc. IEEE.* **56** (1968).
4. Bartelink, D. J. and Scharfetter, D. L. Avalanche shock fronts in p–n junctions. *Appl. Phys. Lett.* **14**, 320, (1969).
5. Carroll, J. E. and Credé, R. H. A computer simulation of TRAPATT circuits. *Int. J. Electronics,* 32, 273 (1972).
6. Clorfeine, A. S., Ikola, R. J. and Napoli, L. S. A theory for the high-efficiency mode of oscillation in avalanche diodes. *R.C.A. Review,* 397, (Sept. 1969).
7. DeLoach, B. C. and Scharfetter, D. L. Device Physics of TRAPATT oscillators. *IEEE Trans. Electron Devices* **ED–17**, 9 (1970).
8. Cottam, M. G. A theory for high-efficiency oscillations in gallium arsenide avalanche diodes, *J. Phys. D.* **3**, 1033 (1970).
9. Cottam, M. G. Theory for high-efficiency oscillations in silicon avalanche diodes. *Electron. Lett.* **6**, 203 (1970).
10. Scharfetter, D. L. Power–frequency characteristic of the TRAPATT diode mode of high-efficiency power generation in germanium and silicon avalanche diodes. *Bell Syst. tech. J.* 799. (May–June 1970).
11. Clorfeine, A. S. Guide lines for the design of high-efficiency mode avalanche diode oscillators. *IEEE Trans. Electron Devices* **ED–18**, 550 (1971).
12. Evans, W. J. and Scharfetter, D. L. Characterization of avalanche diode TRAPATT Oscillators. *IEEE Trans. Electron Devices* **ED–17**, 397 (1970).
13. Gibbons, G. and Grace, M. I. High-efficiency avalanche-diode oscillators and amplifiers in X-band. *Proc. IEEE.* **58**, 512 (1970).
14. Evans, W. J. Computer experiments on TRAPATT diodes. *IEEE Trans. Microwave Theory & Tech.* **17**, 1060 (1969).
15. Evans, W. J., Seidel, T. E. and Scharfetter, D. L. A novel TRAPATT oscillator design. *Proc. IEEE.* **58**, 1284 (1970).
16. Evans, W. J. Circuits for high-efficiency avalanche-diode oscillators. *IEEE Trans. Microwave Theory & Tech.* 17, 1060 (1969).
17. Ying, R. S. and Kramer, N. B. X band silicon TRAPATT diodes. *Proc. IEEE.* **58**, 1285 (1970).
18. Slaymaker, N. A. and Carroll, J. E. Efficient power extraction at TRAPATT harmonics, *Electron. Lett.* **6** (1970).
19. Liu, S. G. Harmonic extraction from high-efficiency avalanche diodes. *Proc. IEEE* **59**, 1216 (1971).
20. Liu, S. G. and Risko, J. J. Fabrication and performance of kilowatt L band avalanche diodes. *R.C.A. Review,* **31**, 3 (1970).
21. Iglesias, D. E. and Evans, W. J. high-efficiency CW IMPATT operation. *Proc. IEEE.* **56**, 1610 (1968).

Index

abrupt-junction diodes, 23, 36, 38, 41, 56, 62, 70, 103, 105, 117, 122
  one-sided, avalanche breakdown in, 30–33
  power–frequency limits, 81–86
accumulation layer, 93–94
acoustic-phonon scattering, 4, 6–8
active region, 26, 30
alloy technique, 90, 91, 94
ambient temperature, 17, 78, 92
anomalous mode, 108
avalanche breakdown,
  condition, 28–30, 68
  one-sided abrupt junction, 30–33
  Read diode, 34–55
  symmetrical step junctions, 33–34
  TRAPATT diodes, 35–38
avalanche current, 48, 78
avalanche delay, 46–48
avalanche multiplication, 17–21, 69
avalanche resonance frequency, 51–53, 64, 108
avalanche shock front, 109–111, 114, 125
  velocity of, 111, 115–118
avalanche zone impedance, 50–51

band structure of GaAs, 11, 12
band-gap energy of semiconductor, 5
  change with temperature, 17
  effect on breakdown field, 30
  effect on power-handling, 106
  effect on reverse saturation current, 78
band-to-band transitions, 32
bandwidth, IMPATT diode, 101
Baraff plots, 15–17
barrier, potential, at metal–semiconductor boundary, 93, 94
bias post, 91, 102
Boltzmann transport equation, 15
breakdown, avalanche; *see* avalanche breakdown
broad-band circuits, 125
burn out of junction, 78, 81

carrier generation, 25, 35, 46, 57
circuits, IMPATT oscillators, 97–103
  millimetre-wave, 102–103
circuits, TRAPATT, 127–129
coaxial circuit, 77, 97, 100, 101
coax-waveguide configuration, 103, 104
communication systems, 2, 105
computer simulations,
  Read diode oscillator, large-signal, 62–64
  TRAPATT oscillators, 125
contact,
  rectifying, 93
  ohmic, 93
contact resistance, 88
contacts, metal, 79, 80, 90, 93–95
construction techniques, IMPATT diodes, 89–92
continuity equations for holes and electrons, 46–47
Conwell–Weisskopf formula, 7
coulomb scattering, 7
current density, critical value of, 55
CW conditions, 78, 103
CW limit, 121
CW operation, 85, 121, 125, 127, 131
CW oscillators, 62, 105, 106, 128, 133
CW output power, 105, 106

density of states in conduction band of semiconductor, 94
depletion layer at metal–semiconductor boundary, 93–94
depletion-layer approximation, 26–28
depletion-layer capacitance, 53
diffused-junction IMPATT, 90
diffusion current, 24–25
displacement current, 23, 51, 75, 109, 111
diode chips, 90, 91, 96, 105, 131
doping concentration,
  effect on breakdown voltage, 31
  effect on field, 23
double-drift diode, 33–34, 37, 56, 78, 97, 105

double-sided diode, 126–127
double-slug circuit, 130
drift velocity, 5–12
saturated, 4, 11
drift-zone impedance, 52–53

effective lifetime for recombination, 25
effective mass of electrons in semiconductor, 6–7, 11–12
effective temperature of electron velocity distribution, 8
efficiency, IMPATT oscillator, 48–49, 61, 66–76, 95, 103–106
abrupt-junction IMPATT, 83
large-signal analysis, 64
space-charge effects, 54
thermal limitations, 78
efficiency, TRAPATT oscillator, 108, 124, 125
electron affinity of semiconductor, 93
electroplating, 89, 90
epitaxy, 75, 76, 88–90, 94, 97, 104, 122
etching techniques, 90, 96
extended-avalance model, 56–57, 67
external load, power transfer to, 74
extraction time, 116–117

feedback effect, 18
Fermi level, 94
field emission of electrons, 32–33
figure of merit, 102
flip-chip mounting, 104
FM noise, 95, 101–102
frequency, IMPATT oscillators, 103–106
frequency, TRAPATT oscillators, 118
frequency modulation, 101–102
fundamental TRAPATT frequency, 125, 128–131

Gauss' law, 53, 55, 57, 64
Gaussian distribution of impurities, 97
growth rate of oscillation, 61, 64
Gunn oscillator, 1, 12, 76, 89

harmonic extraction, 129–131
heat sink, 79, 89–91
copper, 79, 81, 85, 100, 105, 106
diamond, 80–81, 85, 91, 104, 105, 106
high-field effects in semiconductors, 4–21
high-frequency diodes, fabrication techniques, 95–97
high-frequency modulation sensitivity, 102
high-frequency TRAPATTS, 125–127
hole diffusion coefficient, 25
hole–electron pair, creation of, 5, 14
hot electrons, 2, 4, 8

ideal diode equation, 25
impact ionization, 12–17
IMPATT mode, 2
principles of operation, 44–64
IMPATT diodes, series combination, 91–92
IMPATT oscillators, performance limitations, 68–86
imperfections in crystals, electron collisions with, 5–7
impurity diffusion, 88
impurity-scattering limit of mobility, 7
integral heat sink technique, 90
intrinsic carrier concentration, 25
intrinsic region, carrier generation in, 35
ion implantation technique, 97
ionic conduction, 25
ionization energy, 5, 12–13, 17
ionization process, effect of heating on, 17
ionization rates, 12–17
effect on efficiency of IMPATT oscillators, 67–70
saturation effect at high fields, 70–72

junction-side-down mounting, 104, 105

klystron, 2

lattice temperature, 8–9
large-signal analysis, IMPATT oscillator, 62–64
low-pass filter, TRAPATT circuit, 127

matching network, 74, 97, 102–103
mean free path (of electrons in semiconductor), 5
for producing ionization, 12–14
for optical-phonon emission, 12–17
merit, figure of, 102
mesa diodes, 90
millimetre-wave IMPATTS, circuits for, 102–103
minority carrier storage, 72–74
mobility, 5–12
modulation sensitivity, high-frequency, 102
mounting stud, 92, 96, 102
multiple uniform-layer approximation, 56, 60–62
multiplication factor, 18, 20–21, 28

negative resistance, 1, 4–5, 44, 53, 57–58, 78
noise, IMPATT diode, 101–102, 105

Ohm's law, 9
deviations from, 4
one-sided abrupt junctions; *see* abrupt-junction diodes

optical phonons, 10
  thermal excitation, 16
optical-phonon absorption, 16
optical-phonon collisions, 5, 13, 15, 16
  average energy loss, 17
optical-phonon emission, 4, 11, 12
optical-phonon energies, 11, 14, 17, 77
oscillator efficiency; *see* efficiency, oscillator

packaged device, 90, 91
packaging techniques, 96, 102
parametric amplifiers, 2, 105
parasitic capacitance, 96
parasitic inductance, 96
parasitic resistance, 26, 40, 74–76, 88, 89, 96, 104
passivated beam lead devices, 95
performance of IMPATT oscillators, 103–106
  limitations, 66–86
phased-array radar system, 2
plasma, 109
  trapped-plasma formation, 109–111
  trapped-plasma state, 109, 123, 126, 130–131
plasma extraction, 111–113
plasma region, 115–116
plasma velocity, 115–118
Poisson's equation, 27, 40, 56, 113, 116, 118
power of IMPATT oscillator, 103–106
  input, 77, 78
  output, 75, 76, 78, 101
power of TRAPATT oscillators, 108, 131
  power-frequency relation, 118–121, 133
pulsed-bias conditions, 108
punch-through, 89, 122
  diode, 35–39, 117
  factor, 35, 36, 117, 121, 122–123, 125

Q-factor, 61–62, 75, 101–103
quantum-mechanical tunnelling, 25, 33, 94
  Zener tunnelling, 32–33

random noise fluctuations, 44, 99
Read diode, 1, 37, 38, 39
  avalanche breakdown, 34–35
  impedance, 49–53
  large-signal numerical solution, 62–64
  space-charge, 40–41, 53–56
recombination, effective lifetime for, 25
recovery period, 111–113, 122
recovery region, TRAPATT oscillator, 116–117
relaxation time, 5–7
relaxation time for optical-phonon scattering, 11
resistive impedance limitations, 84
resonant frequency, 44, 74, 98, 102
resonant-cap configuration, 102
reverse saturation current; *see* saturation current
reversed-biased p–n junction, 23–42

saturation current, 25, 46, 48, 72–73, 78
saturated drift velocity, 4, 11
Schottky barrier,
  junction, 74, 89
  diode, 106
single-drift diode, 34, 78, 105
skin depth, 84, 96, 102
skin effects, 89
small-signal amplifier, 61
small-signal analysis, 56–62
small-signal impedance, 49–53, 56
  avalanche zone, 50–51
  drift zone, 52–53
space-charge effects, 53–56
space-charge resistance, 40–42, 92–93
spreading resistance in heat sink, 79
stability of IMPATT oscillator circuits, 97–102
stand-offs, 96–97
static breakdown, 48
steady-state current, 6
subharmonic number, 122, 125, 129, 131
substrates, 88–90, 96, 109, 122
surface leakage, 25
surface states of semiconductor, 94, 106
symmetrical step junction,
  potential distribution in, 28
  avalanche breakdown, 33–34

temperature coefficient of breakdown voltage, 41, 92
temperature effects, 40–42
thermal equilibrium, electrons with lattice, 9
thermal impedance of diode, 41, 78–80, 91–93, 104, 121
thermal limitations to performance of IMPATT oscillators, 78–81
thermal time constant, 92–93
thermally generated resistance, 40, 42
thermally limited current density, 85
thermocompression bonding, 96
transferred-electron oscillator, 12
transient mode of breakdown of p–n junction, 109
transistors, 76
transit angle, 53, 60, 62, 122
transit time, 45, 48, 50–51, 53–54

transit-time delay, 1, 44–46
transit-time frequency, 2, 108, 109
transit-time limits, 76–78, 108
TRAPATT diodes, avalanche breakdown
in, 35–38
TRAPATT oscillators, 108–133
trapped-plasma state; *see* plasma
triggering mechanisms, TRAPATT oscillators,
114–115, 126
tunability of device, 101
tuning of circuits,
IMPATT, 97–98
TRAPATT, 127, 130

ultrasonic bonding technique, 91
uniform-flux boundary condition, 80
uniformly avalanching region, 56–59

vacuum evaporation, 90
vapour-deposition technique, 88
varactor diode, 102
velocity of sound in crystal, 9

waveguide circuit, 77, 97, 100, 101, 129
WKB approximation, 33
work function, 93–94

Zener tunnelling, 32–33